Representation and Understanding

Studies in Cognitive Science

LANGUAGE, THOUGHT, AND CULTURE: *Advances in the Study of Cognition*

Under the Editorship of: E. A. HAMMEL

DEPARTMENT OF ANTHROPOLOGY
UNIVERSITY OF CALIFORNIA
BERKELEY

Michael Agar, Ripping and Running: A Formal Ethnography of Urban Heroin Addicts

Brent Berlin, Dennis E. Breedlove, and Peter H. Raven, Principles of Tzeltal Plant Classification: An Introduction to the Botanical Ethnography of a Mayan-Speaking People of Highland Chiapas

Mary Sanches and Ben Blount, Sociocultural Dimensions of Language Use

Daniel G. Bobrow and Allan Collins, Representation and Understanding: Studies in Cognitive Science

In preparation

Domenico Parisi and Francesco Antinucci, Essentials of Grammar

Elizabeth Bates, Language and Context: The Acquisition of Pragmatics

Eugene S. Hunn, Tzeltal Folk Zoology: The Classification of Discontinuities in Nature

Ben G. Blount and Mary Sanches, Sociocultural Dimensions of Language Change

Representation
and Understanding
Studies in Cognitive Science

edited by

DANIEL G. BOBROW
Xerox Palo Alto Research Center
Palo Alto, California

and

ALLAN COLLINS
Bolt Beranek and Newman
Cambridge, Massachusetts

ACADEMIC PRESS, INC.   New York   San Francisco   London   1975

A Subsidiary of Harcourt Brace Jovanovich, Publishers

ACADEMIC PRESS, INC.
111 Fifth Avenue, New York, New York 10003

United Kingdom Edition published by
ACADEMIC PRESS, INC. (LONDON) LTD.
24/28 Oval Road, London NW1

LIBRARY OF CONGRESS CATALOG CARD NUMBER: 75-21630

ISBN 0–12–108550–3

PRINTED IN THE UNITED STATES OF AMERICA

This book is affectionately dedicated to Jaime Carbonell
by his friends and colleagues.

CONTENTS

Contents

PREFACE

Jaime Carbonell was our friend and colleague. For many years he worked with us on problems in Artificial Intelligence, especially on the development of an intelligent instructional system. Jaime directed the Artificial Intelligence group at Bolt, Beranek, and Newman (in Cambridge, Massachusetts) until his death in 1973. Some of us who had worked with Jaime decided to hold a conference in his memory, a conference whose guiding principle would be that Jaime would have enjoyed it. This book is the result of that conference.

Jaime Carbonell's important contribution to cognitive science is best summarized in the title of one of his publications: *AI in CAI*. Jaime wanted to put principles of Artificial Intelligence into Computer-Assisted Instruction (CAI) systems. He dreamed of a system which had a data base of knowledge about a topic matter and general information about language and the principles of tutorial instruction. The system could then pursue a natural tutorial dialog with a student, sometimes following the student's initiative, sometimes taking its own intiative, but always generating its statements and responses in a natural way from its general knowledge. This system contrasts sharply with existing systems for Computer-Assisted Instruction in which a relatively fixed sequence of questions and possible responses have to be determined for each topic. Jaime did construct working versions of his dream--in a system which he called SCHOLAR. But he died before SCHOLAR reached the full realization of the dream.

It was a pleasure to work with Jaime. His kindness and his enthusiasm were infectious, and the discussions we had with him over the years were a great stimulus to our own thinking. Both as a friend and a colleague we miss him greatly.

Cognitive Science. This book contains studies in a new field we call *cognitive science*. Cognitive science includes elements of psychology, computer science, linguistics, philosophy, and education, but it is more than the intersection of these disciplines. Their integration has produced a new set of tools for dealing with a broad range of questions. In recent years, the interactions among the workers in these fields has led to exciting new developments in our understanding of intelligent systems and the development of a science of cognition. The group of workers has pursued problems that did not appear to be solvable from within any single discipline. It is too early to predict the future course of this new interaction, but the work to date has been stimulating and inspiring. It is our hope that this book can serve as an illustration of the type of problems that can be approached through interdisciplinary cooperation. The participants in this book (and at the conference) represent the fields of Artificial Intelligence, Linguistics, and Psychology, all of whom work on similar problems but with different viewpoints. The book focuses on the common problems, hopefully acting as a way of bringing these issues to the attention of all workers in those fields related to cognitive science.

Subject Matter. The book contains four sections. In the first section, **Theory of Representation**, general issues involved in building representations of knowledge are explored. Daniel G. Bobrow proposes that solutions to a set of design issues be used as dimensions for comparing different representations, and he examines different forms such solutions might take. William A. Woods explores problems in representing natural-language statements in semantic networks, illustrating difficult theoretical issues by examples. Joseph D. Becker is concerned with the representation one can infer for behavioral systems whose internal workings cannot be observed directly, and he considers the interconnection of useful concepts such as hierarchical organization, system goals, and resource conflicts. Robert J. Bobrow and John Seely Brown present a model for an expert understander which can take a collection of data describing some situation, synthesize a *contingent knowledge structure* which places the input data in the context of a larger structural organization, and which answers questions about the situation based only on the contingent knowledge structure.

Section two, **New Memory Models**, discusses the implications of the assumption that input information is always interpreted in terms of large structural units derived from experience. Daniel G. Bobrow and Donald A. Norman postulate active *schemata* in memory which refer to each other through use of *context-dependent descriptions,* and which respond both to input data and to hypotheses about structure. Benjamin J. Kuipers describes the concept of a *frame* as a structural organizing unit for data elements, and he discusses the use of these units in the context of a recognition system. Terry Winograd explores issues involved in the controversy on representing knowledge in declarative versus procedural form. Winograd uses the concept of a frame as a basis for the synthesis of the declarative and procedural approaches. The frame provides an organizing structure on which to attach both declarative and procedural information.

The third section, **Higher Level Structures,** focuses on the representation of plans, episodes, and stories within memory. David E. Rumelhart proposes a grammar for well-formed stories. His summarization rules for stories based on this grammar seem to provide reasonable predictions of human behavior. Roger C. Schank postulates that in understanding paragraphs, the reader fills in causal connections between propositions, and that such causally linked chains are the basis for most human memory organization. Robert P. Abelson defines a notation in which to describe the intended effects of plans, and to express the conditions necessary for achieving desired states.

The fourth section, **Semantic Knowledge in Understander Systems,** describes how knowledge has been used in existing systems. John Seely Brown and Richard R. Burton describe a system which uses multiple representations to achieve expertise in teaching a student about debugging electronic circuits. Bonnie Nash-Webber describes the role played by semantics in the understanding of continuous speech in a limited domain of discourse. Allan Collins, Eleanor H. Warnock, Nelleke Aiello, and Mark L. Miller describe a continuation of work on Jaime Carbonell's SCHOLAR system. They examine how humans use strategies to find reasonable answers to questions for which they do not have the knowledge to answer with certainty, and how people can be taught to reason this way.

Preface

Acknowledgments. We are grateful for the help of a large number of people who made the conference and this book possible. The conference participants, not all of whom are represented in this book, created an atmosphere in which interdisciplinary exploration became a joy. The people attending were:

From Bolt Beranek and Newman--Joe Becker, Rusty Bobrow, John Brown, Allan Collins, Bill Merriam, Bonnie Nash-Webber, Eleanor Warnock, and Bill Woods.

From Xerox Palo Alto Research Center--Dan Bobrow, Ron Kaplan, Sharon Kaufman, Julie Lustig, and Terry Winograd (also from Stanford University).

From the University of California, San Diego--Don Norman and Dave Rumelhart. From the University of Texas--Bob Simmons. From Yale University--Bob Abelson. From Uppsala University--Eric Sandewall.

Julie Lustig made all the arrangements for the conference at Pajaro Dunes, and was largely responsible for making it a comfortable atmosphere in which to discuss some very difficult technical issues. Carol Van Jepmond was responsible for typing, editing, and formatting the manuscripts to meet the specifications of the systems used in the production of this book. It is thanks to her skill and effort that the book looks as beautiful as it does. June Stein did the final copy editing, made general corrections, and gave many valuable suggestions on format and layout.

Photo-ready copy was produced with the aid of experimental formatting, illustration, and printing systems built at the Xerox Palo Alto Research Center. We would like to thank Matt Heiler, Ron Kaplan, Ben Kuipers, William Newman, Ron Rider, Bob Sproull, and Larry Tesler for their help in making photo-ready production of this book possible. We are grateful to the Computer Science Laboratory of the Xerox Palo Alto Research Center for making available the experimental facilities and for its continuing support.

Representation
and Understanding

Studies in Cognitive Science

DIMENSIONS OF REPRESENTATION

Daniel G. Bobrow
Xerox Palo Alto Research Center
Palo Alto, California

I. INTRODUCTION

Workers in cognitive science have worried about what people know, and how to represent such knowledge within a theory.[1] Psychologists such as Paivio (1974) and Pylyshyn (1973) have argued, for example, over two alternative forms for visual memory in humans. The style of their arguments, which we return to at the end of this chapter, is to set up opposing characterizations and to argue about which one has more "natural" properties with respect to observed phenomena.

I claim that a more appropriate way of discussing the issues involved is to characterize each representation in terms of how it answers certain questions posed in this chapter. I pose these questions in terms of a set of design issues one would face in designing or analyzing an *understander system*--a system (human or computer) which could use the knowledge to achieve some goal. I propose a framework for viewing the problems of representation. In this framework each of the design issues defines a dimension of representation--a relatively independent way of looking at representations.

In this chapter I emphasize the structure of alternative solutions to the design issues. I illustrate the design options through three specific representations described here, and in examples from the literature and other chapters in this book. By considering representations along the separate dimensions, it often becomes apparent that a pair of seemingly disparate representations differ in very few significant features.

A. Representation and Mapping

I propose here a framework where representations are viewed as the result of a selective mapping of aspects of the world. Suppose we take a "snapshot" of the world in a particular state at some instant in time. Call this state *world-state-1*. Through some mapping M, a representation

---

[1]In the preface we describe cognitive science as a new field containing elements from psychology, linguistics, computer science, philosophy, education, and artificial intelligence.

(call it *knowledge-state-1*) is created which corresponds to world-state-1. This corresponds with world-state-1 in the sense that an understander has the alternative of answering questions about world-state-1 by directly observing the world state or by questioning the corresponding knowledge state (see Fig. 1).

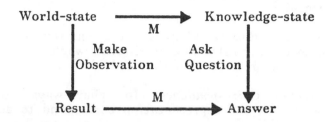

Fig. 1. Mapping between world and knowledge states. Answering questions should correspond to making observations and mapping the result.

This implies, of course, the existence of a world-observation and knowledge-question function correspondence; simplicity of the mapping M, and simplicity of representing particular knowledge and questions must be considered in comparing representations of a world-state.

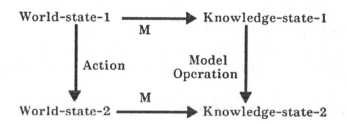

Fig. 2. A world-state can be changed by an action. An equivalent model operation should produce a change in the knowledge-state which corresponds to the changed world-state.

The world at a particular instant is static, and all the facts about the world reflect a single consistent state. If we now augment our simple view, and allow *actions* which change some properties of the world, then we must have *model operations* which make corresponding changes in the knowledge state. For a model to be consistent, an updated world-state-2 must correspond to the updated knowledge-state-2 (see Fig. 2).

In terms of this simple framework for viewing representation, we can now look at a number of different design issues. I pose these as a series of questions to be asked about any mapping and the resulting representation of the world:[2]

Domain and Range: What is being represented? How do objects and relationships in the world correspond to units and relations in the model?

Operational Correspondence: In what ways do the operations in the representation correspond to actions in the world?

Process of Mapping: How can knowledge in the system be used in the process of mapping?

Inference: How can facts be added to the knowledge state without further input from the world?

Access: How are units and structures linked to provide access to appropriate facts?

Matching: How are two structures compared for equality and similarity?

Self-awareness: What knowledge does a system have explicitly about its own structure and operation?

B. Three Simple Visual Representations

To illustrate some options concretely on certain dimensions, I use three different specific representations for the same simple domain--two-dimensional black and white scenes. I describe how each represents a visual scene which contains a square rotated so that one diagonal is horizontal.

------

[2]In constructing this list of questions, I have been influenced by the dimensional analysis used by Moore & Newell (1973) in describing their system MERLIN.

Binary Matrix: Fig. 3a shows a two-dimensional binary matrix represention (MATRIX) of the spatial layout. A "1" is inserted in the matrix wherever the light intensity in the scene is below some threshold, and a "0" otherwise.

```
0 0 0 0 0 0 0 0 0 0
0 0 0 0 1 0 0 0 0 0
0 0 0 1 1 1 0 0 0 0
0 0 1 1 1 1 1 0 0 0
0 1 1 1 1 1 1 1 0 0
0 0 1 1 1 1 1 0 0 0
0 0 0 1 1 1 0 0 0 0
0 0 0 0 1 0 0 0 0 0
0 0 0 0 0 0 0 0 0 0
```

Fig. 3a. A binary matrix visual representation.
A 1 indicates a light intensity below a certain level.

A collection of connected 1s determines an object, with transitions between spaces containing 1s and 0s indicating the contours of an object.

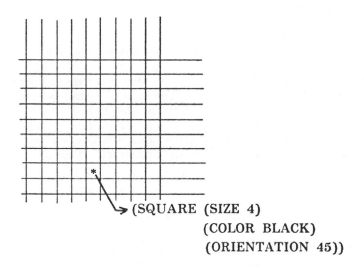

(SQUARE (SIZE 4)
(COLOR BLACK)
(ORIENTATION 45))

Fig. 3b. A Grid-positioned/feature oriented representation.

Grid-positioned feature: Fig. 3b shows what I call a grid-positioned feature representation (GRID) for a scene. An object is represented by a unit which specifies a set of features. The structure shown is of type SQUARE, with features specifying the size, color, and orientation of the square. The definition of SQUARE is not shown; it can be obtained given the type specification. The grid is used to locate objects in a scene. From a point on the grid corresponding to the location of the leftmost lowest point of the object, there is a link to the unit representing the object.

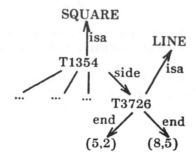

Fig. 3c. A semantic network representation.

Semantic Network: Fig. 3c shows a portion of a semantic network represention (NET) of the same visual scene. The units shown are a token of a square, tokens of sides of the square, and some number pairs representing the endpoints of the sides of the square. Only one of the endpoint sets are shown. Labelled links from one unit to another show the relations between the units.

II. DOMAIN AND RANGE

A. Units and Relations

The choice of units and structures reflects how one views the world one is modeling. A unit is something which can be used without knowing anything about its internal structure. This does not imply necessarily that it must not have any internal structure, just that there are occasions of use (e.g., inference rules) in which the existence of the unit is sufficient. In addition to its identity, response to a unit may be a function of its position in a larger containing structure, or special relation to other units, or to its internal form.

In choosing a representation for a particular world, some relationships can be stored explicitly and others need not be. For example, the size of the square is implicit in the matrix in MATRIX, as is the position of the square in GRID. These are reflections of what Hayes (1974) describes as the similarity between the medium of representation and the world, at least with respect to the relations being modeled.

Not all relations in a representation fully determine a portion of the world. For example, the relative position of two objects (A is left of B) may be implicit in locations represented in the model. Alternatively, this fact may be explicitly represented, with perhaps no absolute location information for either unit. How such "vague" predicates and partial information about the world are handled is an important characteristic of a representation. (See Chapter 2 by Woods for a more complete discussion of problems of vague predicates.)

B. Exhaustiveness

A representation is exhaustive with respect to a property if for any object, if it has that property, that fact is stored explicitly. Not only does the model represent the truth, it represents the whole truth. Thus in an exhaustive representation of the objects present on the surface of a table top, any object not explicitly noted as on top is not there, and if no object is associated with a location, then that location is guaranteed to be empty. In an exhaustive representation all objects that exist are represented explicitly, and any universal proposition can be verified by testing all elements of this set of objects. Exhaustiveness is a second aspect of what Hayes (1974) refers to as similarity of structure of the medium.

One way for a visual representation to maintain the property of exhaustiveness is for the mapping to have the property of extracting a uniform degree of detail. An aerial photographer does map terrain this way whereas a cartographer may not. In the photo, it is guaranteed that no object within the field of view and larger than the resolution of the lens will be missing. The whim of the mapmaker determines the objects and features represented on a map. MATRIX is by nature exhaustive; GRID and NET can be made so by design. Human visual memory does not seem to have this property of uniform extraction of detail, or of exhaustiveness.

C. Verbal Mediation

Instead of mapping the world directly, people have constructed systems which map the world using natural language descriptions. There are many issues involved in building adequate representations of English language statements. Woods (Chapter 2) points, for example, to the subtle problems involved in representing relative clauses and verbal restrictions within a semantic network. In this chapter, I focus only on the issues involving selection of basic units to represent linguistic information.

Word-Senses: An obvious choice for a unit is a single word. The relations chosen often are the case relations for verbs (Fillmore, 1968). This simplifies the mapping process by focusing on the obvious units and their grammatical relationships. A problem with this choice is that words are often ambiguous. Some systems finesse this issue by assuming that each word will have only one meaning within the domain of interest. Other systems face the issue by allowing individual words a number of different senses.

Several problems must be considered in systems which use word-senses. There is the obvious potential error in ignoring concepts for which there are no single words (or for which the user knows none), such as a single word to describe "those small orange cones used to divert traffic". A word-sense system must allow compound constructs to be used as well as atomic units.

Another problem arises if the system is forced to make an either/or decision, since use of the word may straddle two word-senses, even though word-senses in the dictionary are usually chosen so that they are distinguishably far apart. For example, consider the word "weigh" in "The butcher weighed the meat", which has a different sense than the same word in "The jury weighed the evidence." Instead of interpreting these sentences in terms of separate meanings for "weigh", we can consider its common core-- "weigh" as "comparing an unknown with a contextually determined scale". If selection of a word-sense precludes using any part of any other meaning, then intermediate creative uses can be missed, particularly those arising in metaphorical use of language.

A system which uses word-senses must also provide a way of determining the equivalence of two different phrases. The sentences:

John sold the boat to Bill.
Bill bought the boat from John.

have identical factual meanings, although the difference in topicalization may be used to guide the storage of the information in long-term memory. Recognition of the equivalence of these paraphrases for purposes of inference requires translation rules to convert from one form into another, or separate inference rules for each form [see Chapter 2, and Simmons (1973) for a more complete discussion of this issue].

Semantic Primitives: If simplification of the paraphrase problem is made a focus of the representation design, the mapping can be designed to translate all input to a canonical form, so that identity of meaning is equivalent to identity of form. One way to do this is to expand all input to expressions involving only a small fixed set of primitive units as the basic relations. Schank (Chapter 9) gives a number of arguments for use of such an expansion, and describes a set of eleven ACTs (Schank's primitive units) which are useful for this purpose. All actions in his system are expressed in terms of this primitive set, and the predicates define states which result from these actions.

Schank claims that this representation has the additional advantage that inferences which should be made when a new sentence comes in can be keyed to the individual ACTs rather than having to be stored with each word or word-sense.

As an argument against expansion to primitives, one notes that there are significant inferences which must be made on the basis of particular combinations of ACTs and states for a particular word. Consider the differences in obvious inferences between these two statements:

John thought something, which caused him to do something, which caused a male actor to become in a state of worst possible health.

John killed him for a reason.

This is only a mild caricature of expansion to primitives. The point is, to recognize the situation from the longer paraphrase, a more complex match is required than for the more compact one. Thus there is a tradeoff between the types of operations that can be done easily in the two representations. This is really a form of the tradeoff of processing at input time versus processing at time of use of information.

III. OPERATIONAL CORRESPONDENCE

One issue of major concern in representation is the correspondence between action and structures in the physical world, and operations and representational forms in a model. Simple actions in the domain should be reflected in simple operations in the representation.

A. Updating and Consistency

A major design problem in modeling actions is updating the representation with respect to a chain of changes caused by a single action. For example, suppose at time t1 a cup of coffee is on a table. At time t2 the cup is pushed over the edge of the table, and at time t3 all has settled down again after the actions starting at t2. It is easy to see how a system might represent world-state-1. It is easy to see how the action at time t2 might be represented. The problem comes in determining how a representation might reflect the facts that if a cup falls off a table, the cup has changed position, but the table has not and the contents of the cup (the coffee) is no longer either on the table or in the cup. How does the system determine which of the facts true at t1 are true at t3? Must every model operation have associated with it a way of checking for all possible implications? Or should relations and objects in the world be represented by active entities which check for conditions affecting them? Small systems have been built on each assumption, and the tradeoffs are just beginning to be explored.

A complementary problem exists if not all operations in the model correspond directly to actions in the world; a unitary operation may correspond to only one element of an action. The representation may then allow knowledge-states to be constructed which cannot be realized in the world. For example, in GRID, the extent of each object is not indicated on the grid. Therefore, if in the knowledge state an object is moved so that its attachment point is adjacent to another object's attachment point, there is no obvious violation, but the objects may overlap unacceptably in the real world.

B. History and Planning

In the simple mapping of static world-states, time is represented only implicitly in terms of changes in the world, and corresponding changes in the knowledge base. A more sophisticated representation would allow the simultaneous representation of two different world-states, so that a history can be stored. A major problem is how to find and represent shared pieces of the two knowledge states. Problems related to shared structure have been discussed extensively in the artificial intelligence literature using the phrase "the frame problem", for example by Raphael (1971). Unfortunately, this use of the word "frame" differs in meaning from the term "frame" used in this book.

In the previous section on updating and consistency I assumed the need for only one set of "true" facts, true for the time the world-state was modeled. If a history is to be stored, then those facts which *were* true but are not now, must be distinguished. One alternative is to associate with every fact time bounds for the truth of that fact. Sometimes this is difficult in that the start or end times may not be known. An alternative proposed by Sandewall (1971) is to mark changes as they occur. A fact is assumed true at any time t2 if it was true at some earlier time, t1, and the system cannot prove that the fact has become untrue. Thus facts need not be repeated for each instant, but it may be costly to check on the truth of a fact.

A related problem occurs in planning. Planning is a search for a series of actions to bring about a particular desired world-state. In conducting the search, shared knowledge and updating problems must be dealt with. In planning, the changes are not real; they result from modeling activity, not world activity. A search can be made by actually carrying out a plausible sequence of operations and testing the result. Exploring the search space of operations is a difficult task, and needs to be guided by common sense. Abelson (Chapter 10) discusses a formalism in which common sense reasoning about plans can be embedded.

In order to allow backup in case of error, a copy of the original world state must be kept, or provision must be made to allow "undoing" operations (reversing their effects in the model). In addition, if two worlds are to be compared, there must be some way to save information about alternative hypothetical states of the world. An obvious possibility is to have multiple worlds in which operations are applied to separate copies of the knowledge or modeling base. Here the system must provide some way of knowing which information is shared and which is unshared.

C. Continuity

As we have defined them, operations change the knowledge-state to reflect the differences in the world-state at the beginning and endpoints of an action. One argument for certain types of models concerns the intermediate states the model should go through when certain operations are performed. Model operations can be implemented so as to permit only small incremental changes in the model.

For example, in any of the visual representations, one could imagine that a large change of location by an object could only be made as a sequence of small changes in the representation. Thus in MATRIX, rotations and translations might be made by moving one cell of the matrix one unit at a time. In GRID, rotation might add only a small increment to the orientation value, perhaps limited by the resolution with which such information is specified. In either case, if one wished to model a reasonably large movement, the object representation would have to proceed incrementally through space from the initial position to the desired position, perhaps checking in the model for interpenetration of objects on the path. If operations had enforced continuity, other constraints on the trajectory could also be checked, as well as possible transient effects such as blocking of light.

D. Psychological Modeling

If a purpose of the representation is to provide a psychological model of some mental activity, then some correspondence must be defined between measurable resources used by a person and invocation of some operations in the model. If the obvious choices were made, then in MATRIX the time for mental rotation of an object would be proportional to the area of the object (number of 1s in the matrix). In NET, it would be proportional to the number of endpoints of lines; and for GRID it would be independent of the shape and size.

Imposing continuity on an operation like rotation allows the assumption that the change in the representation takes time proportional to distance traveled. This, of course, is analogous to transformations in the real world, despite the fact that the underlying format of the representational model might not normally be considered analogous to the world. Models with this property for humans seem to be implied by the data of Cooper & Shepard (1973). In their experiments, subject were asked to compare a rotated figure with a possibly identical one in a standard position. The time to make the comparison was a linear function of the amount of rotation, but independent of the complexity of the line drawings used (Cooper 1975).

IV. THE MAPPING PROCESS

A. Constraints on World States

In our discussion thus far it has been assumed that units and relations correspond to particular objects and relationships in the world. If knowledge in the model is to help in the mapping, some facts must represent constraints over sets of world states (perhaps even all mappable states). In order to do this a system can use formulas or structures containing *variables*. One can interpret such a structure as follows: if appropriate constants are substituted for the variables in the structure, then the relationship expressed will be true.

Mapping design issues then center around mechanisms for specifying and finding constants which satisfy the appropriate constraints; this is related to the issue of determining the size and structure of units retrieved and used in the mapping process.

Restrictions on Variables: In addition to satisfying the relations indicated in formulas, variables are often subject to other restrictions. A simple restriction is on the type of the entity which can be substituted for a variable. For example, the value of an attribute COLOR can be specified to be one of the color-names. In some cases, restriction on a range of values is also specified; for example, a day of the month must be an integer and between 1 and 31. Restrictions on variables can be used to check possible substitutions, or can be used to help in a search for objects in the world which are appropriate.

Other restrictions, aside from range, can arise from interaction of selection of variable substitutions. In one reading of the statement "every professor X loves some student Y", then the choice of Y is dictated by the identity of the element which is chosen for X. This is typical of issues which arise in quantification. It is here that formal representations based on the predicate calculus have advantages, in that the subtleties of the conections have been worked out. Woods (Chapter 2) gives examples of a number of problems in quantification as expressed in English, and works out forms of representation for quantification for use in a semantic network.

Higher Level Structures and Mapping: Predicate calculus and semantic network representations tend to impose only a local organization on the world. Much of the thrust of research reported in this book deals with organizing information at higher structural levels such as frames (Chapters 6 and 7), scripts (Chapters 9 and 10), and story frameworks (Chapters 8 and 9). Such higher-level structures help in the description and instantiation of structures as complicated as birthday parties or a complete story. Default values are often provided for variables in such structures so that a priori guesses can be used to provide a complete picture with little or no processing. Such guesses would most often be right (a person usually has two legs), and need only be verified by a quick test. Kuipers (Chapter 6) gives an example using a frame for a standard clock, in which an object identified as a clock is represented as having hands, although this was assumed from default, and in the example was an incorrect assumption.

Larger structures provide a conceptual framework on which to hang inputs. By forcing inputs to fit within an expected framework, however, the system will see only what it wants to. Alternatively, using the data to drive processing will cause extensive search as numerous low-level combinations of units will be found which cannot be used in larger structures. Bobrow & Norman (Chapter 5) discuss tradeoffs between such concept-driven and data-driven processing, suggesting both are needed, and must interact.

B. Procedural Declarative Tradeoffs

Information used in processing can be isolated in declarative data structures or can be embedded in procedures for achieving special purposes. In the extreme, one can take the view that "knowing" is "knowing how to" and that all behavior is engraved in programs. Hewitt (1971) has been a major proponent of expressing all knowledge as procedures. The issues between declarative and procedural underlying structures are extensively discussed by Winograd (Chapter 7), so I only summarize the arguments here. Declarative languages provide economy of representation (many uses for the same knowledge), and human understandability and communicability. They rely on general procedures using special problem-dependent data. Procedural embedding emphasizes use of specific procedures for specific problems, allowing easy use of control information (second order knowledge) and easy representation of process-oriented information. Winograd characterizes these features in terms of modularity versus interaction, and makes a preliminary proposal for joining declarative and procedural representations.

V. INFERENCE

Not all of the facts in any knowledge-state need be kept explicitly in a representation. If partial knowledge is available in the system, then some set of explicit facts may have implications, that is, determine further facts which satisfy the constraints represented in the particular knowledge state. *Inference* is the process of deriving implicit facts from the initial set of explicit formulas according to some fixed rules of inference without interaction with the outside world. The form of inferences available and the structure of data to support these inferences are important design decisions for a system.

I distinguish among three different forms of inference. The term *formal inference* covers the family of techniques used in predicate calculus representational systems. The term *computational inference* describes a process in which facts are derived through bounded known computation. The term *meta-inference* covers techniques by which knowledge about the structure and content of the data base is used to derive further facts consistent with the original set. A major forcing function in design of a representation can be the desire to make a particular set of inferences easy. These *preferred inferences* are often what give a representation much of its power.

A. Formal Inference Techniques

In formal systems, facts about the world are represented in quantified formulas. To infer a new explicit fact one produces a formal proof, with a chain of intermediate formulas produced through modus ponens, resolution, or another standard syntactic method. All the facts which support an inference are thus available for inspection, in contrast to computational inference discussed in the next section.

A formal basis for inference has a number of strong implications for the representation chosen. McCarthy & Hayes (1969) have pressed a strong case for using a formal (predicate calculus) substrate for an understanding system. Advantages cited include the use of theorem proving techniques which are not domain-dependent. Thus these techniques are applicable whether the information represented concerns trip planning, children's stories, or the physical world of robots. Formal systems often have the property of completeness; that is, the proof techniques guarantee that if a fact *can* be proved from those available to the system, then, given enough time, that fact *will* be proved to be true. The logic of quantifiers and their interaction has been extensively worked out, and the predicate calculus takes advantage of this long history of careful thought.

The other side of each of these arguments is as follows: If only general theorem proving is used, then special facts about the domain (for example, classifications of facts as useful in particular types of inferences) are made difficult or impossible to use. The property of completeness is often not really useful because the condition "given enough time and space" is often unfulfillable. Moreover the system is built on the assumption that it contains only a consistent set of facts. If what are being represented is the beliefs of an individual at some time, then this set of beliefs may indeed not be consistent, or at least not expressed in a consistent manner. For example, the generalization that "all birds can fly" can usefully live in a system which contains the specific facts "ostriches cannot fly" and "an ostrich is a bird". In classical logic, the existence of these three inconsistent facts would allow the deduction of any arbitrary fact.

Two techniques are often used to work around the problem of contradictory beliefs in the same data base. An ordering principle can be used, where specific facts are considered before generalizations. Alternatively, the quantifier "all" can be reinterpreted to mean "all, unless I tell you about an exception". In either case, formal properties of the quantifiers disappear, and standard proof procedures can no longer be used.

B. Computational Inference

In some systems a specialized procedure is used for computing certain facts about the state of the world. For example, for a world consisting of a set of objects placed on a two-dimensional plane, there may be a list of that set of objects and their x-y coordinates. Additional facts can be derived using *procedural specialists*, such as a "left-of" specialist that uses the coordinate information to answer the question about whether the object X is to the left of object Y. In SOPHIE (Chapter 11), procedural specialists use the contingent voltage table and a resistance-connection table for the circuit to answer questions about currents and power dissipation in any element in the circuit.

There are two points to make to distinguish this form of inference from the formal techniques. First, control information--which facts to use next and how--is built into the procedure. Therefore search procedures are not included in this class of computations. Although this is often efficient, it can lead to some rigidity in how the procedure works and which parameters it can use. The idea is, by specialization to a particular assumed environment, special case data and control tests can be avoided.

Second, no intermediate results are available as in formal inference; no justification of any result is given. For example, the input to the circuit simulator of Brown & Burton (Chapter 11) are values of circuit elements. A relaxation method--make an approximation, find the errors, try again--determines a consistent set of voltages across elements. The result is accepted as valid without proof; the only guarantee of correctness is at the level of initially proving the program correct (or just debugging it).

This works if all inputs are valid, and no anomalous cases occur. If errors occur, a simple procedural model can only throw up its hands. One alternative, used in PLANNER (Hewitt, 1971), is to have a set of specialists for doing individual tasks; if one specialist fails, the system reverts to its earlier state (backtracks) and tries another specialist. In PLANNER, desired inferences are classified by their syntactic form and content. Procedures to make such inferences are invoked on the basis of the form of the fact to be proved. All inferences in PLANNER are carried out by procedural specialists written by the user.

Sussman (1973) has built a program for debugging programs which is based on computational inference. The procedural steps in the program to be debugged are augmented by a set of "intention statements" which can be checked against the program for various known forms of "bugs" (errors in programming). Winograd (Chapter 7) describes a system with a procedural base, which also contains a declarative description of procedures that are invoked. If the compiled version (the one which is unitary and leaves no trace) runs into unexpected problems, then the task can be rerun in a more careful mode using the procedure description.

C. Meta-Inferential Techniques

Some systems have been designed to find facts which are not necessarily derivable in a formal way from the set already present, but which are consistent with such a set and may be useful.

Inductive Inference: One class of techniques, inductive inference, uses a set of facts to form the basis for a general rule for expressing relations. The general rule is consistent with the given data but may not necessarily be correct; it may be later contradicted by additional data. Using a general rule to replace specific data can save space in information storage. It may also provide a basis for new theorems in the system. Brown (1973) discusses a number of problems in building an inductive inference system which occur even in a very limited world; one of the worst problems is the existence of faulty data.

Inference by Analogy: Another class of inference techniques goes under the general rubric of inference by analogy. In inference by analogy, if certain criteria of similarity are met between two situations, then a result that pertains to the first situation can be assumed to pertain to the second situation. Collins, Warnock, Aiello, & Miller (Chapter 13) discuss a particular form of this inference by analogy in SCHOLAR which they call *functional inference.* They distinguish major conditions they call *functional determinants* which are critical in allowing a geographical location to have a particular property. For example, the latitude and altitude of a place are the major functional determinants of the type of climate at that place. SCHOLAR uses the following rule for inference by analogy.

> If a property P has functional determinants F and G, and F and G are identical for place 1 and place 2, then barring information to the contrary, if place 1 has property P, then assume place 2 has that property as well.

Thus since Los Angeles and Sydney, Australia are both at sea level and at 33 degrees latitude, their climates should be similar.

Learning this type of functional knowledge is an important part of human learning in general, and such functional rules allow one to generate many reasonable answers without formally sufficient data. In general, for an analogic inference, the criterial properties of a situation with respect to some result must be marked and stored in the representation.

Self Knowledge Inferences: Another class of meta-inferences taken from the SCHOLAR program (Chapter 13) is based on the system's knowledge of its own internal structure. SCHOLAR uses information about the importance of particular properties, and level of relevance of facts it has about a particular place. This extension of the exhaustiveness property discussed earlier allows determination of negative answers based on not finding information in the data base; without such knowledge the system would often be forced to reply "I don't know".

As a simple example, consider how the SCHOLAR program answers the question, "Is oil a major product of Chile?" SCHOLAR knows that copper is a major product of Chile, and oil is a major product of Venezuela. It also knows facts which are less important than the major products of Chile, so it assumes it knows all products that are of major importance. It thus responds, "No, oil is probably not a major product of Chile."

As another example, consider a question discussed by Norman (1973), "What is Charles Dickens' phone number?" Most humans (and hopefully most intelligent systems) will be able to answer "I don't know" immediately without having to do a long search. Again this is based on knowing what is known, and how easily accessible such knowledge is.[3] Norman proposes multiple stages of search, with an initial filtering done on the basis of knowledge of the system's own knowledge.

D. Preferred Inferences

Each system has certain inferences which can be made more easily than others. Often this is designed into a

---

[3]A Chas. Dickens is listed in the Palo Alto phone book; I looked it up after Allen Newell asked me how I knew there wouldn't be such a listing (as he found out) in the Pittsburgh directory. If a person knew a Charles Dickens other than the well-known author, this fact would probably be unusual enough to make it immediately accessible when such a question were asked; thus, with this assumption the question could again be answered with assurance.

system. For example, in most semantic nets a preferred
inference attributes to an individual any property of the
general class to which it belongs. For example, Fido would
inherit all the properties of a generic dog, e.g., he has four
legs, he barks, etc. These preferred inferences often give a
system much of its power; this has certainly been true with
semantic networks (Quillian, 1969).

Some systems derive preferred inferences on the basis of
examples. That is, for a particular set of inputs, they
derive an example which satisfies the inputs, and then
check to see whether a suggested subgoal is true in the
example. From general specification of a geometric figure,
Gelernter's (1960) geometry machine drew (inside the
computer) an example of a figure satisfying the
specification. Any constructions and any hypotheses
considered were first checked against this particular
instance to see if they were reasonable. Brown & Burton
(Chapter 11) discuss another way of using examples.

VI. ACCESS

Use of the appropriate piece of knowledge at the right
time is the essence of *intelligent* mental operations. Two
different issues arise in the consideration of access to data
and procedures. The first concerns the philosophy of which
elements to link together. The second concerns mechanisms
which are used for access.

Since access and storage are inverse operations, there is
a tradeoff between work done at each time. I assume that
retrieval (access) is done significantly more often than
storage; therefore I focus only on the access issue, and
assume for this exposition that any necessary work has
been done to allow the access regimes discussed.

A. Philosophy of Association

In each system there are implicit access links between
elements of a single structure, and links which join
structures. The former reflects which things in the world
are viewed as unitary structures. The latter is used to

facilitate internal processing such as making inferences. In predicate calculus representations, the natural structure is the formula. A single formula contains a number of different relations among units. The relation is the critical item, and so organizational aspects of the structure are based around selection of the relations. It is made easy to determine which relations have been used together, and harder to find all the potential properties of an individual, and the relations in which it has appeared.

In semantic networks, the organizational aspects of objects are emphasized, and the relations appear primarily in the interconnection between the units. A semantic network makes immediately accessible the kinds of relations that an individual participates in. It is possible to test whether information about a known individual is new or redundant, inconsistent or derivable from previous relations. Use of variables, and constraints between variables are harder to represent. Schank (Chapter 9) and Woods (Chapter 2) discuss alternatives in structuring semantic nets for access.

In more direct models, such as GRID or MATRIX, access is usually defined in terms of spatial location. Near neighbors of a point are directly accessible, and properties of that point are easily available; for example, the contents of that point can be found immediately.

One way of building structures larger than single formulas is to consider contexts in which relations are used. If a particular set of facts, or network structures, are used for understanding a particular situation, then that entire context can usefully be retrieved at one time. Those contexts themselves may be organized into still higher-level contexts. For example, the meaning of the words "cost" and "buyer" may best be understood in the context of knowledge about commercial transactions; further implications will come from a context generally applicable to a monetary economic system.

Another possible organizational structure is in terms of scenarios. Here, some higher-level structure which one wants to impose on the world is used to tie together otherwise disparate facts. The problem of putting together the individual structures of a representation in terms of higher-order structures has been a major thrust of a number of chapters in this book. Schank (Chapter 9) places causal links between propositions in a paragraph; Rumelhart (Chapter 8) proposes a structure which describes well-formed stories; and Abelson (Chapter 10) develops a language for describing plans. Winograd (Chapter 6) deals with the issue of associating specialized procedures with frames. The frame not only acts as an organizational structure for data, but for procedures as well.

B. Access Mechanisms

Suppose a unit of information is placed in a known location, for example starting at some address A in a computer. Then this information is directly accessible (without search) to a process which knows A. That is, there is an operation basic to the computer which retrieves the contents of a cell given the address of that cell. The address A can be stored as an element in another unit of information (call it B). If B contains A, we say that B has a pointer to A. This is how semantic networks represent links between elements. One of the main features of semantic networks is this explicit representation of interconnections between memory units.

If direct access is not available, then a *retrieval* mechanism must be invoked which will take a description of a desired unit, and search memory for a unit which fits the description. Ordering and structuring memory can speed up a retrieval search. The cost is paid at storage time, either in placement of items or in updating indexes.

It is important to note that usually only part of a description (a "key") is used in the search, and then the potential candidate is *matched* against the full description to determine its appropriateness. Sometimes, although a single direct pointer is not given, a list specifying a set of possibilities is provided. Then a description is used only for checking the possible candidates.

The description of a unit to be accessed can be constructed from both stored and dynamic information. For example, current context can delimit the set of possible elements which are of interest, and only a brief description need be used to discriminate one of these. Pronominal reference in English makes use of such assumed context for successful operation. Bobrow & Norman (Chapter 5) make a case for context-dependent descriptions being the primary basis for access in an intelligent system.

An access mechanism which is much discussed, but which has not yet been used in any artificial intelligence systems, is an active content addressable memory. In such a memory, a description of a desired unit would be broadcast to many (perhaps all) active memory units. Each would compare its own contents with the request, and answer if a good enough match were found. Problems which must be faced include specification of how good the match must be, how to get the information back to the requester, how to deal with conflicts, and how to resolve timing problems if more than one request is active at a time. Because of present hardware and software limitations, such a system has not been tried, although procedural systems such as CONNIVER (Sussman, 1972), have used software to simulate some of the properties. A goal pattern for a procedure is specified, and an access mechanism invokes procedures which have been stored with a "trigger" pattern which matches the goal pattern. Bobrow & Raphael (1974) describe this pattern-directed invocation, and a number of other properties of the new artificial intelligence programming languages.

Another access mechanism which can use an active memory system is the "intersection" technique simulated in many semantic network models. Here access is specified in terms of two key elements which are both to be associated by a chain of direct links with the desired item. From each of these keys the network is searched by following pointers from each key, in a breadth first fashion, until an element is reached by search from one key which has previously been reached from the other. More than one intersection may be found if parallel active search is going on. Models using this type of access have been proposed for human processing. Collins & Quillian (1972) among others have conducted a number of interesting experiments which give some evidence for this type of search in human language processing.

VII. MATCHING

A. Uses for Matching

Matching as an operation can be used for a number of purposes within an intelligent system: *classification, confirmation, decomposition* and *correction.* To determine the identity of an unknown input, a number of possible labeled patterns can be matched against the unknown. The unknown can be *classified* in terms of the pattern it matches best. This is the paradigm for simple pattern recognition. In retrieval, a possible candidate to fit a description may be *confirmed* by the match procedure. If it matches well enough, the retrieval and match together provide a pattern-directed access capability. A pattern with substructure can be matched against a structured unknown, and the unknown *decomposed* into subparts corresponding to those in the pattern. A parsing system is a complicated pattern matcher whose purpose is to find substructures corresponding to patterns in the grammar. In certain matches, what is critical is the form or direction of the error in the match. In hill climbing or relaxation techniques, a first approximation to a solution is *corrected* by use of this error term. Kuipers (Chapter 6) discusses using errors of prediction as a guide in pattern recognition.

B. Forms of Matching

Systems frequently use matching for one or more of the above purposes, and purpose can be confounded with the form of matching done in the system; we describe three basic forms of matching, syntactic, parametric, and semantic, and a mode of forced matching.

Syntactic Matches: In syntactic matching, the form of one unit is compared with the form of another, and the two forms must be identical. In a slight generalization, a unit may have variables, which can match any constant in the other. Further complications involve putting restrictions on the types of constants a variable can match. A common use of syntactic matching procedures is to find appropriate substructures by matching variables which fit into parts of larger structures. Another step in the generalization is to allow the pattern matcher to be recursive, so that the matcher is called to determine if a subpiece of a pattern matches a subpiece of a unit. Bobrow & Raphael (1974) describe classes of variable restrictions and pattern matching in current AI programming languages.

Parametric Matches: In syntactic match, a binary decision is made. A pattern either does or does not match. In a parametric match, a parameter specifies the goodness of any match. In such a match, certain features of a pattern may be considered essential, others typical and hence probably should be there, and others just desirable in an element to be matched. A goodness parameter can account for how many of which features can be found. Ripps, Shoben & Smith (1974) hypothesize that people use a parametric match using levels of feature comparison. For example, they claim a person would classify a particular picture of an animal as a bird if sufficient features presented in that picture match those of a "typical" bird.

Semantic Matches: In a semantic match, the form of elements are not specified. The function of each element in the structure is specified; then the system must engage in a problem-solving process to find elements which can serve that function. For example, a table could be specified to be a horizontal surface on top of a support which keeps the surface at a height of about 30 inches. This does not at all specify the form of the support, which could be anything from a box to a cantilever from a wall. This type of specification, separating form from function, seems necessary to allow the flexible definitions that humans seem capable of handling.

Forced Matches: Moore & Newell (1974) in their MERLIN system, discuss a mapping process in which one structure is viewed as though it were another. Matches of corresponding items in the structures are forced if necessary. Forcing such matches allows certain operations applicable to one unit to be used in conjunction with the other. For example, if you were in a locked room and wished to get out, you could break open the window if you had a hammer. If no hammer were available, it might occur to you to view your shoe as a hammer; the sole would be forced to match the handle of the hammer, and the heel the head. Bobrow & Norman (Chapter 5) discuss procedures for building in the generalizability required by such forced matches by using minimal descriptions. These descriptions serve to aid in the identification of relevant matches, and to handle the necessary applications of the constraints on these variables.

VIII. SELF-AWARENESS

An important dimension of system design is whether the system has explicit knowledge of its own workings. This dimension has not been well explored in representation systems, and so I give here only a menu of different kinds of self-awareness that might be built into a system.

A. Knowledge about Facts

Exhaustiveness of a representation with respect to a property is a form of self-knowledge which we discussed with respect to operational correspondence and meta-inference. It is generalizable to the level of relevance, as in the SCHOLAR system. A related property is a level of importance or interest associated with classes of facts. This type of knowledge is useful in forward inferencing schemes in which resources have to be allocated; inferences based on interesting or important new facts should be made first.

Criteriality is a term used to describe the relevance of the identity or truth of some element in a match. Becker (1973) uses the adjustment of criterialities as a basis for automatic generalization of experience. Another class of knowledge about facts concerns the belief status of a fact. Values of belief between true and false can be used, as well as the basis on which the belief was acquired. For example, a system may remember that it was told a particular fact by Richard, and therefore it is much less likely to be true. Providing criteriality or expected degree of validity of information is important when a contradiction is encountered. This is the type of knowledge that a system must have in order for it to be able to correct errors in its own procedures. Other useful facts about facts are characterizations of situations in which they are useful. Classification of the kinds of facts known, and the importance for different functions is a level of self awareness whose utility we describe in the section on inference.

B. Knowledge about Process

In modeling interactions with the outside world, the system needs to predict its own capabilities to plan a strategy in which information gathering cannot all be done before starting an action sequence. For example, in planning a route, it must be able to realize that at a certain intersection it will be able to look for a street sign.

Other process-knowledge information is relevant to a system which has different strategies for solving problems with special characteristics. Characterization of a problem should be a first step in deciding when to apply domain-specific heuristics. Information useful for scheduling competing processes is important in multigoal systems. Such knowledge includes resource requirements for procedures, and a priori and dynamic estimates of success of particular problem solving routines. I believe all of these levels of self-awareness will be necessary for us to build intelligent understander systems.

IX. CONCLUSION

A. Multiple Representations

It is often convenient (and sometimes necessary) to use several different representations within a single system. In this way, it is sometimes possible to combine the advantages of different representational forms within one system. The use of multiple representations leads to two primary problems: choice and consistency.

Choice: In a system with multiple representations, a particular fact may be represented in several ways. In such cases, a system must contain a mechanism to choose which form to use for any particular fact. For example, the location of an object may be given by its coordinates, or in terms of its relative location to some other object, or it may be placed in a grid with the location implicit in the grid point on which it is located. If different representations are used, mechanisms are needed to transform information in one representation to that of another. Sometimes, however, the information in one representation does not allow a reasonable transformation to another. For example, knowing that A is to the left of B does not position A precisely enough to allow it to be placed on a grid, even if B is on that grid.

Representations determine the ease of answering certain questions, and of performing updating operations. At times, it is best to enter information directly into one representational form and then, from there, compute how to enter it into the other form. Thus an object's position might first be entered by its coordinates, and then its position relative to all others computed and inserted into the appropriate representation. Questions about its relative position and its absolute position then can be answered with equal facility.

Consistency: In a system with multiple representations the same information can be stored in more than one form. When one form changes, the other forms must be checked for consistency. For example, if the left-right relations of an object have been stored, and the object is moved, all those relations must be recomputed. An alternative is to maintain a primary data representation, such as the positional information. Secondary information can be represented in procedural form, with special procedures to compute the desired results quickly.

Updating is a more serious problem in representations in which facts may have been inferred on the basis of a large number of other facts. The multiple representation problem compounds the problems of single representation consistency, updating, and planning I discussed in Section III.

Efficiency: Major considerations in use of multiple representations are tradeoffs between computation and storage, and availability of special techniques for achieving efficiency; for a particular process all information may be transformed to the preferred representation. For example, in Chapter 11 by Brown & Burton, a dual representation system is used for electronic circuits. Circuit calculations are performed by a circuit simulator which provides descriptions of particular, consistent states of the circuit. The simulator implicitly embeds in its operations knowledge of the interactions and feedback among circuit elements. A semantic network in their system, which stores propositional information, is excellent for answering many types of questions; but it would founder on the feedback issue.

B. Analog representations

In psychology, a current debate rages over how visual information is represented in human memory--whether or not it is stored in "analog" form. For example, Sloman (1971) points to implicit interaction as an important argument for analogic representations. Pylyshyn (1973) argues that if information is stored as images it must have a uniform degree of detail in the representation. Given the known fine detail a person can sometimes store, uniform extraction implies an overload of information in picture memory. Paivio (1974) rejects the uniform-detail position in arguing for images. He claims, however, that propositional models can not have appropriate continuity in operations, thus failing to model the Cooper-Shepard results I described in IIID. This characterization over-simplifies the arguments, but indicates the dimensional nature of the disagreements.

Having representational dichotomies such as *analogical* versus *propositional* requires, I think, that we make overly sharp distinctions. In this chapter, I illustrated the properties inherent in a choice of representation for visual scenes by discussing three possibilities: MATRIX, NET, and GRID. The units of MATRIX are only the visual elements, and relative location is an implicit relation between two units. NET has named symbols as units, and named relations linking them, with no implicit relations. GRID has two types of units, grid points to record positions, and symbolic units represented in a list of property value pairs at some of the grid points. Only MATRIX is "naturally" exhaustive, though the other two can be made so explicitly. Whereas MATRIX seems "obviously" analogical, and NET propositional, it is harder to decide about GRID. I believe that such debate is best viewed by considering claims made along separate dimensions.

The most distinguishing feature of these representations is along the dimension of access. Properties of a point are directly accessible from the location in MATRIX. In NET such information can only be found by search and computation. Access to a unit as an entity is direct in GRID and NET, and requires a search in MATRIX. In GRID, but not in NET, one can access a square (or any unit) directly knowing its center of mass. In NET, but not GRID, the coordinates of the corners are a directly accessible property of an object. I did not define properties of MATRIX, GRID, and NET with respect to operational correspondence, mapping process, inference, matching or self-awareness. Often in an isolated model, significant differences in theory rest on which dimensions of representation are *not* considered.

This chapter provides the reader with a set of questions to ask about representations of knowledge. The questions are organized in terms of a mapping framework, with dimensions corresponding to design issues which must be faced or finessed in any representation. It is my experience that viewing representations along these multiple dimensions allows more complete and coherent evaluations and comparisons.

ACKNOWLEDGMENTS

I would like to thank Don Norman without whose help and encouragement this chapter would never have been written. I am also grateful to the following people for comments which greatly improved the final result: Rusty Bobrow, Lynn Cooper, Ron Kaplan, Martin Kay, Allen Newell, and Terry Winograd.

REFERENCES

Becker, J. D. A model for the encoding of experiential information. In R. C. Schank & K. M. Colby (Eds.), *Computer models of thought and language.* San Francisco, Ca.: Freeman, 1973.

Bobrow, D. G., & Raphael, B. New programming languages for artificial intelligence research. *Computing Surveys,* 1974, *6(3),* 153-174.

Brown, J. S. Steps toward automatic theory formation. *Third International Joint Conference on Artificial Intelligence.* Stanford University, August 1973, 121-129.

Collins, A. M., & Quillian, M. R. How to make a language user. In E. Tulving and W. Donaldson (Eds.), *Organization of memory.* New York: Academic Press, 1972.

Cooper, L. A. Mental rotation of random two-dimensional shapes, *Cognitive Psychology,* 1975, 1, 20-43.

Cooper, L. A., & Shepard, R. N. Chronometric studies of the rotation of mental images. In W. G. Chase (Ed.), *Visual information processing.* New York: Academic Press, 1973.

Fillmore, C. J. The case for case. In Bach & Harms (Eds.), *Universals in linguistic theory.* Chicago, Ill.: Holt, 1968.

Gelernter, H. Realization of a geometry theorem proving machine. *Proceedings of 1959 International Conference on Information Processing.,* 1960, 273-282.

Hayes, P. Some problems and non-problems in representation theory. *Proceedings of the A.S.B Summer Conference, Essex University,* 1974, 63-79.

Hewitt, C. Description and theoretical analysis (using schemata) of PLANNER: A language for proving theorems and manipulating models in a robot. Ph.D. Thesis (June 1971) (Reprinted in AI-TR-258 MIT-AI Laboratory, April 1972.)

McCarthy, J., & Hayes, P. Some philosophical problems from the standpoint of artificial intelligence. In Meltzer and Michie (Eds.), *Machine intelligence 4.* Edinburgh University Press, 1969.

Moore, J., & Newell, A. How can MERLIN understand?. In Gregg (Ed.), *Knowledge and cognition*. Baltimore, Md.: Lawrence Erlbaum Associates, 1973.

Norman, D. A. Memory, knowledge, and the answering of questions. In R.L. Solso (Ed.), *Contemporary issues in cognitive psychology: The Loyola symposium*. Washington, DC: Winston, 1973.

Norman, D. A., & Bobrow, D. G. On data-limited and resource-limited processes. *Cognitive Psychology*, 1975, *7*, 44-64.

Norman, D. A., Rumelhart, D. E., & the LNR Research Group. *Explorations in cognition*. San Francisco: Freeman, 1975.

Paivio, A., Images, propositions, and knowledge. Research Bulletin No. 309. London, Canada: Department of Psychology, The University of Western Ontario, 1974.

Pylyshyn, Z. W. What the mind's eye tells the mind's brain: a critique of mental imagery. *Psychological Bulletin*, 1973, *80*, 1-24.

Quillian, M. R. The teachable language comprehender: A simulation program and theory of language. *Communications of the ACM*, 1969, *12*, 459-476.

Raphael, B. The frame problem in problem-solving systems. In *AI and heuristic programming*. Edinburgh University Press, 1971.

Rips, L. J., Shoben, E. J., & Smith, E. Structure and process in semantic memory: A featural model for semantic decisions. *Psychological Review*, 1974, *81(3)*, 214-241.

Sandewall, E. Representing natural language information in predicate calculus. *Machine Intelligence 6*. Edinburgh University Press, 1971.

Simmons, R. F. Semantic networks: Their computation and use for understanding English sentences. In R. C. Schank & K. M. Colby (Eds.), *Computer models of thought and language*. San Francisco, Ca.: Freeman, 1973.

Sloman, A. Interactions between philosophy and artificial intelligence. *Artificial Intelligence 2*, 1971.

Sussman, G. J. A computational model of skill acquisition. MIT-AI Laboratory AI TR-297 (August 1973).

Sussman, G., & McDermott, D. From PLANNER to CONNIVER - A genetic approach. Fall Joint Computer Conference. Montvale, N. J.: AFIPS Press, 1972.

WHAT'S IN A LINK:

Foundations for Semantic Networks

William A. Woods
Bolt Beranek and Newman
Cambridge, Massachusetts

I. INTRODUCTION

This chapter is concerned with the theoretical underpinnings for semantic network representations of the sort dealt with by Quillian (1968,1969), Rumelhart, Lindsay, & Norman (1972), Carbonell & Collins (1973), Schank (1975), Simmons (1973), etc. (I include Schank's conceptual dependency representations in this class although he himself may deny the kinship.) I am concerned specifically with understanding the semantics of the semantic network structures themselves, i.e., with what the notations and structures used in a semantic network can mean, and with interpretations of what these links mean that will be logically adequate to the job of representing knowledge. I want to focus on several issues: the meaning of "semantics", the need for explicit understanding of the intended meanings for various types of arcs and links, the need for careful thought in choosing conventions for representing facts as assemblages of arcs and nodes, and several specific difficult problems in knowledge representation--especially problems of relative clauses and quantification.

I think we must begin with the realization that there is currently no "theory" of semantic networks. The notion of semantic networks is for the most part an attractive notion which has yet to be proven. Even the question of what networks have to do with semantics is one which takes some answering. I am convinced that there is real value to the work that is being done in semantic network representations and that there is much to be learned from it. I feel, however, that the major discoveries are yet to be made and what is currently being done is not really understood. In this chapter I would like to make a start at such an understanding.

I will attempt to show that when the semantics of the notations are made clear, many of the techniques used in existing semantic networks are inadequate for representing knowledge in general. By means of examples, I will argue that if semantic networks are to be used as a representation for storing human verbal knowledge, then they must include mechanisms for representing propositions without commitment to asserting their truth or belief. Also they must be able to represent various types of intensional objects without commitment to their existence in the external world, their external distinctness, or their completeness in covering all of the objects which are presumed to exist. I will discuss the problems of representing restrictive relative clauses and argue that a commonly used "solution" is inadequate. I will also demonstrate the inadequacy of certain commonly used techniques which purport to handle quantificational information in semantic networks. Three adequate mechanisms will be presented, one of which to my knowledge has not previously been used in semantic nets. I will discuss several different possible uses of links and some of the different types of nodes and links which are required in a semantic network if it is to serve as a medium for representing knowledge.

The emphasis of this chapter will be on problems, possible solution techniques, and necessary characteristics of solutions, with particular emphasis on pointing out nonsolutions. No attempt will be made to formulate a complete specification of an adequate semantic network notation. Rather, the discussion will be oriented toward requirements for an adequate notation and the kind of explicit understanding of what one intends his notations to mean that are required to investigate such questions.

II. WHAT IS SEMANTICS?

First we must come to grips with the term "semantics". What do semantic networks have to do with semantics? What is semantics anyway? There is a great deal of misunderstanding on this point among computational linguists and psychologists. There are people who maintain that there is no distinction between syntax and semantics, and there are others who lump the entire inference and "thought" component of an AI system under the label "semantics". Moreover, the philosophers, linguists, and programming language theorists have notions of semantics which are distinct from each other and from many of the notions of computational linguists and psychologists.

What I will present first is my view of the way that the term "semantics" has come to be associated with so many different kinds of things, and the basic unity that I think it is all about. I will attempt to show that the source of many confusing claims such as "there is no difference between syntax and semantics" arise from a limited view of the total role of semantics in language.

A. The Philosopher and the Linguist

In my account of semantics, I will use some caricatured stereotypes to represent different points of view which have been expressed in the literature or seem to be implied. I will not attempt to tie specific persons to particular points of view since I may thereby make the error of misinterpreting some author. Instead, I will simply set up the stereotype as a possible point of view which someone might take, and proceed from there.

First, let me set up two caricatures which I will call the Linguist and the Philosopher, without thereby asserting that all linguists fall into the first category or philosophers in the second. Both, however, represent strong traditions in their respective fields. The Linguist has the following view of semantics in linguistics: he is interested in characterizing the fact that the same sentence can sometimes mean different things, and some sentences mean nothing at all. He would like to find some notation in which to express the different things which a sentence can mean and some procedure for determining whether a sentence is "anomalous" (i.e., has no meanings). The Philosopher on the other hand is concerned with specifying the meaning of a formal notation rather than a natural language. (Again, this is not true of all philosophers--just our caricature.) His notation is already unambiguous. What he is concerned with is determining when an expression in the notation is a "true" proposition (in some appropriate formal sense of truth) and when it is false. (Related questions are when it can be said to be necessarily true or necessarily false or logically true or logically false, etc.) Meaning for the Philosopher is not defined in terms of some other notation in which to represent different possible interpretations of a sentence, but he is interested in the conditions for truth of an already formal representation.

Clearly, these caricatured points of view are both parts of a larger view of the semantic interpretation of natural language. The Linguist is concerned with the translation of natural languages into formal representations of their meanings, while the Philosopher is interested in the meanings of such representations. One cannot really have a complete semantic specification of a natural language unless both of these tasks have been accomplished. I will, however, go further and point out that there is a consideration which the philosophers have not yet covered and which must be included in order to provide a complete semantic specification.

B. Procedural Semantics

While the types of semantic theories that have been formulated by logicians and philosophers do a reasonable job of specifying the semantics of complex constructions involving quantification and combination of predicates with operators of conjunction and negation, they fall down on the specification of the semantics of the basic "atomic" propositions consisting of a predicate and specifications of its arguments--for example, the specification of the meanings of elementary statements such as "snow is white" or "Socrates is mortal". In most accounts, these are presumed to have "truth conditions" which determine those possible worlds in which they are true and those in which they are false, but how does one specify those truth conditions? In order for an intelligent entity to know the meaning of such sentences it must be the case that it has stored somehow an effective set of criteria for deciding in a given possible world whether such a sentence is true or false. Thus it is not sufficient merely to say that the meaning of a sentence is a set of truth conditions--one must be able to specify the truth conditions for particular sentences. Most philosophers have not faced this issue for atomic sentences such as "snow is white."

Elsewhere I have argued (Woods, 1967, 1973a) that a specification of truth conditions can be made by means of a procedure or function which assigns truth values to propositions in particular possible worlds. Such procedures for determining truth or falsity are the basis for what I have called "procedural semantics" (although this interpretation of the term may differ slightly from that which is intended by other people who have since used it). This notion has served as the basis of several computer question-answering systems (Woods, Kaplan, & Nash-Webber, 1972; Woods, 1973b; Winograd, 1972).

The case presented above is a gross oversimplification of what is actually required for an adequate procedural specification of the semantics of natural language. There are strong reasons which dictate that the best one can expect to have is a partial function which assigns true in some cases, false in some cases, and fails to assign either true or false in others. There are also cases where the procedures require historical data which is not normally available and therefore cannot be directly executed. In these cases their behavior must be predicted on the basis of more complex inference techniques. Some of these issues are discussed more fully by Woods (1973a).

C. Semantic Specification of Natural Language

You now have the basics of my case for a broader view of the role of semantics in natural language. The outline of the picture goes like this:

There must be a notation for representing the meanings of sentences inside the brain (of humans or other intellects) that is not merely a direct encoding of the English word sequence. This must be so, since (among other reasons) what we understand by sentences usually includes the disambiguation of certain syntactic and semantic ambiguities present in the sentence itself.

The linguist is largely concerned with the process for getting from the external sentence to this internal representation (a process referred to as "semantic interpretation"). The philosopher is concerned with the rules of correspondence between expressions in such notations and truth and falsity (or correctness of assertion) in the real or in hypothetical worlds. Philosophers, however, have generally stopped short of trying to actually specify the truth conditions of the basic atomic propositions in their systems, dealing mainly with the specification of the meanings of complex expressions in terms of the meanings of elementary ones. Researchers in artificial intelligence are faced with the need to specify the semantics of elementary propositions as well as complex ones and are moreover required to put to the test the assembly of the entire system into a working total --including the interface to syntax and the subsequent inference and "thought" processes. Thus the researcher in artificial intelligence must take a more global view of the semantics of language than either the linguist or the philosopher has taken in the past. The same, I think, is true of psychologists.

D. Misconceptions about Semantics

There are two misconceptions of what semantics is about (or at least misuses of the term) which are rather widely circulated among computational linguists and which arise I think from a limited view of the role of semantics in language. They arise from traditional uses of the term which, through specialized application, eventually lose sight of what semantics is really about. According to my dictionary, semantics is "the scientific study of the relations between signs or symbols and what they denote or mean". This is the traditional use of the term and represents the common thread which links the different concerns discussed previously. Notice that the term does not refer to the things denoted or the meanings, but to the *relations* between these things and the linguistic expressions which denote them.

One common misuse of the term "semantics" in the fields of computational linguistics and artificial intelligence is to extend the coverage of the term not only to this relation between linguistic form and meaning, but to all of the retrieval and inference capabilities of the system. This misuse arises since for many tasks in language processing, the use of semantic information necessarily involves not only the determination of the object denoted, but also some inference about that object. In absence of a good name for this further inference process, terms such as "semantic inferences" have come to be used for the entire process. It is easy then to start incorrectly referring to the entire thought process as "semantics". One may properly use the term "semantic inferences" to refer to inferences that cross the boundary between symbol and referent, but one should keep in mind that this does not imply that all steps of the process are "semantic".

At the opposite extreme, there are those who deny any difference in principle between syntax and semantics and claim that the distinction is arbitrary. Again, the misconception arises from a limited view of the role of semantics. When semantics is used to select among different possible parsings of a sentence by using selectional restrictions on so-called semantic features of words, there is little difference between the techniques usually used and those used for checking syntactic features. In another paper (Woods, 1973a) I make the case that such techniques are merely approximations of the types of inferences that are really required, and that, in general, semantic selectional restrictions need to determine the referent of a phrase and then make inferences about that referent (i.e., they involve semantic inferences as I defined the term above). The approximate technique usually used, however, requires no special mechanism beyond what already exists in the syntax specification, and when taken as the paradigm for "semantic inferences" can lead to the false conclusion that semantics is no different from syntax. Likewise, if the representation constructed by a parser purports to be a semantic representation, with no intervening purely syntactic representation, then one might argue that the techniques used to produce it are syntactic techniques and therefore there is nothing left to be semantics.

As we have pointed out, however, a semantic specification requires more than the transformation of the input sentence into a "semantic" representation. The meanings of these representations must be specified also. Recall that semantics refers to the correspondence between linguistic expressions and the things that they denote or mean. Thus although it may be difficult to isolate exactly what part of a system is semantics, any system which understands sentences and carries out appropriate actions in response to them is somehow completing this connection. For systems which do not extend beyond the production of a so-called semantic representation, there may or may not be a semantic component included, and the justification for calling something semantic may be lost. Again, if one takes the production of such "semantic" representations as the paradigm case for what semantics is, one is misunderstanding the meaning of the term.

E. Semantics of Programming Languages

Before proceeding it is probably worth pointing out that the use of the term "semantics" by programming language theorists has been much closer to the tradition of the logicians and the philosophers and less confused than in computational linguistics. Programming language theory is frequently used as a paradigm for natural language semantics. Programming languages, however, do not have many of the features that natural languages do and the mechanisms developed there are not sufficient for modeling the semantics of natural language without considerable stretching.

The programming language theorists do have one advantage over the philosophers and linguists in that their semantic specifications stand on firmer ground since they are defined in terms of the procedures that the machine is to carry out. It is this same advantage which the notion of procedural semantics and artificial intelligence brings to the specification of the semantics of natural language. Although in ordinary natural language not every sentence is overtly dealing with procedures to be executed, it is possible nevertheless to use the notion of procedures as a means of specifying the truth conditions of declarative statements as well as the intended meaning of questions and commands. One thus picks up the semantic chain from the philosophers at the level of truth conditions and completes it to the level of formal specifications of procedures. These can in turn be characterized by their operations on real machines and can be thereby anchored to physics. (Notice that the notion of procedure shares with the notion of meaning that elusive quality of being impossible to present except by means of alternative representations. The procedure itself is something abstract which is instantiated whenever someone carries out the procedure, but otherwise, all one has when it is not being executed is some representation of it.)

III. SEMANTICS AND SEMANTIC NETWORKS

Having established a framework for understanding what we mean by semantics, let us now proceed to see how semantic networks fit into the picture. Semantic networks presumably are candidates for the role of internal semantic representation--i.e., the notation used to store knowledge inside the head. Their competitors for this role are formal logics such as the predicate calculus, and various representations such as Lakoff-type deep structures, and Fillmore-type case representations. (The case representations shade off almost imperceptibly into certain possible semantic network representations and hence it is probably not fruitful to draw any clear distinction.) The major characteristic of the semantic networks that distinguishes them from other candidates is the characteristic notion of a link or pointer which connects individual facts into a total structure.

A semantic network attempts to combine in a single mechanism the ability not only to store factual knowledge but also to model the associative connections exhibited by humans which make certain items of information accessible from certain others. It is possible presumably to model these two aspects with two separate mechanisms such as for example, a list of the facts expressed in the predicate calculus or some such representation, together with an index of associative connections which link facts together. Semantic network representations attempt instead to produce a single representation which by virtue of the way in which it represents facts (i.e., by assemblies of pointers to other facts) automatically provides the appropriate associative connections. One should keep in mind that the assumption that such a representation is possible is merely an item of faith, an unproven hypothesis used as the basis of the methodology. It is entirely conceivable that no such single representation is possible.

A. Requirements for a Semantic Representation

When one tries to devise a notation or a language for semantic representation, one is seeking a representation which will precisely, formally, and unambiguously represent any particular interpretation that a human listener may place on a sentence. We will refer to this as "logical adequacy" of a semantic representation. There are two other requirements of a good semantic representation beyond the requirement of logical adequacy. One is that there must be an algorithm or procedure for translating the original sentence into this representation and the other is that there must be algorithms which can make use of this representation for the subsequent inferences and deductions that the human or machine must perform on them. Thus one is seeking a representation which facilitates translation and subsequent intelligent processing, in addition to providing a notation for expressing any particular interpretation of a sentence.

B. The Canonical Form Myth

Before continuing, let me mention one thing which semantic networks should not be expected to do: that is to provide a "canonical form" in which all paraphrases of a given proposition are reduced to a single standard (or canonical) form. It is true that humans seem to reduce input sentences into some different internal form that does not preserve all of the information about the form in which the sentence was received (e.g., whether it was in the active or the passive). A canonical form, however, requires a great deal more than this. A canonical form requires that *every* expression equivalent to a given one can be reduced to a single form by means of an effective procedure, so that tests of equivalence between descriptions can be reduced to the testing of identity of canonical form. I will make two points. The first is that it is unlikely that there could be a canonical form for English, and the second is that for independent reasons, in order to duplicate human behavior in paraphrasing, one would still need all of the inferential machinery that canonical forms attempt to avoid.

Consider first the motivation for wanting a canonical form. Given a system of expressions in some notation (in this case English, or more specifically an internal semantic representation of English) and given a set of equivalence-preserving transformations (such as paraphrasing or logical equivalence transformations) which map one expression into an equivalent expression, two expressions are said to be equivalent if one can be transformed into the other by some sequence of these equivalence transformations. If one wanted to determine if two expressions $e1$ and $e2$ were equivalent, one would expect to have to search for a sequence of transformations that would produce one from the other--a search which could be nondeterministic and expensive to carry out. A canonical form for the system is a computable function c which transforms any expression e into a unique equivalent expression $c(e)$ such that for any two expressions $e1$ and $e2$, $e1$ is equivalent to $e2$, if and only if $c(e1)$ is equal to $c(e2)$. With such a function, one can avoid the combinatoric search for an equivalence chain connecting the two expressions and merely compute the corresponding canonical forms and compare them for identity. Thus a canonical form provides an improvement in efficiency over having to search for an equivalence chain for each individual case (assuming that the function c is efficiently computable).

A canonical form function is, however, a very special function, and it is not necessarily the case for a given system of expressions and equivalence transformations that there is such a function. It can be shown for certain formal systems [such as the word problem for semigroups (Davis, 1958)] that there can be no computable canonical form function with the above properties. That is, in order to determine the equivalence of a particular pair of expressions $e1$ and $e2$ it may be necessary to actually search for a chain of equivalence transformations that connects these two particular expressions, rather than performing separate transformations $c(e1)$ and $c(e2)$ (both of which know exactly where to stop) and then compare these resulting expressions for identity. If this can be the case for formal systems as simple as semigroups, it would be foolhardy to assume lightly that there is a canonical form for something as complex as English paraphrasing.

Now, for the second point. Quite aside from the possibility of having a canonical form function for English, I will attempt to argue that one still needs to be able to search for individual chains of inference between pairs of expressions *e1* and *e2* and thus the principal motivation for wanting a canonical form is superfluous. The point is that in most cases where one is interested in some paraphrase behavior, the paraphrase desired is not one of full logical equivalence, but only of implication in one direction. For example, one is interested in whether the truth of some expression *e1* is implied by some stored expression *e2*. If one had a canonical form function, then one could store only canonical forms in the data base and ask simply whether *c(e1)* is stored in the data base without having to apply any equivalence transformations in the process. This is, however, just a special case. It is rather unlikely that what we have in the data base is an expression exactly logically equivalent to *e1* (i.e., some *e2* such that *e2* implies *e1* and *e1* implies *e2*). Rather, what we expect in the typical case is that we will find some *e2* that implies *e1* but not vice versa. For this case, we must be able to find an inference chain as part of our retrieval process. Given that we must devise an appropriate inferential retrieval process for dealing with this case (which is the more common), the special case of full equivalence will fall out as a consequence; thus the canonical form mechanism for handling the full equivalence case gives no improvement in performance and is unnecessary.

There is still benefit from "partially canonicalizing" the stored knowledge (the term is reminiscent of the concept of being just a little bit pregnant). This is useful to avoid storing multiple equivalent representations of the same fact. There is, however, little motivation for making sure that this form does in fact reduce all equivalent expressions to the same form (and as I said before, there is every reason to believe that this may be impossible).

Another argument against the expectation of a canonical form solution to the equivalence problem comes from the following situation. Consider the kinship relations program of Lindsay (1963). The basic domain of discourse of the system is family relationships such as mother, father, brother, sister, etc. The data structure chosen is a

logically minimal representation of a family unit consisting
of a male and female parent and some number of offspring.
Concepts such as aunt, uncle, and brother-in-law are not
represented explicitly in the structure but are rather
implicit in the structure and questions about unclehood are
answered by checking brothers of the father and brothers
of the mother. What does such a system do, however, when
it encounters the input "Harry is John's uncle"? It does
not know whether to assign Harry as a sibling of John's
father or his mother. Lindsay had no good solution for
this problem other than the suggestion to somehow make
both entries and connect them together with some kind of
a connection which indicates that one of them is wrong. It
seems that for handling "vague" predicates such as uncle,
i.e., predicates which are not specific with respect to some
of the details of an underlying representation, we must
make provision for storing such predicates directly (i.e., in
terms of a concept of uncle in this case), even though this
concept may be defined in terms of more "basic"
relationships (ignoring here the issue that there may be no
objective criterion for selecting any particular set of
relationships as basic).

If we hope to be able to store information at the level
of detail that it may be presented to us in English, then
we are compelled to surrender the assumptions of logical
minimality in our internal representation and provide for
storing such redundant concepts as "uncle" directly. We
would not, however, like to have to store all such facts
redundantly. That is, given a Lindsay-type data base of
family units, we would not want to be compelled to store
explicitly all of the instances of unclehood that could be
inferred from the basic family units. If we were to carry
such a program to its logical conclusion, we would have to
store explicitly all of the possible inferable relations, a
practical impossibility since in many cases the number of
such inferables is effectively infinite. Hence the internal
structure which we desire must have some instances of
unclehood stored directly and others left to be deduced
from more basic family relationships, thus demolishing any
hope of a canonical form representation.

C. Semantics of Semantic Network Notations

When I create a node in a network or when I establish a link of some type between two nodes, I am building up a representation of something in a notation. The question that I will be concerned with in the remainder of this chapter is what do I mean by this representation. For example, if I create a node and establish two links from it, one labeled SUPERC and pointing to the "concept" TELEPHONE and another labeled MOD and pointing to the "concept" BLACK, what do I mean this node to represent? Do I intend it to stand for the "concept" of a black telephone, or perhaps I mean it to assert a relationship between the concepts of telephone and blackness--i.e., that telephones are black (all telephones?, some telephones?). When one devises a semantic network notation, it is necessary not only to specify the types of nodes and links that can be used and the rules for their possible combinations (the syntax of the network notation) but also to specify the import of the various types of links and structures--what is meant by them (the semantics of the network notation).

D. Intensions and Extensions

To begin, I would like to raise the distinction between intension and extension, a distinction that has been variously referred to as the difference between sense and reference, meaning and denotation, and various other pairs of terms. Basically a predicate such as the English word "red" has associated with it two possible conceptual things which could be related to its meaning in the intuitive sense. One of these is the set of all red things--this is called the *extension* of the predicate. The other concept is an abstract entity which in some sense characterizes what it *means* to be red, it is the notion of *redness* which may or may not be true of a given object; this is called the *intension* of the predicate. In many philosophical theories the intension of a predicate is identified with an abstract function which applies to possible worlds and assigns to any such world a set of extensional objects (e.g., the intension

of "red" would assign to each possible world a set of red
things). In such a theory, when one wants to refer to the
concept of redness, what is denoted is this abstract
function.

E. The Need for Intensional Representation

The following quotation from Quine (1961) relating an
example of Frege should illustrate the kind of thing that I
am trying to distinguish as an internal intensional entity:

> The phrase "Evening Star" names a certain large
> physical object of spherical form, which is hurtling
> through space some scores of millions of miles from
> here. The phrase "Morning Star" names the same
> thing, as was probably first established by some
> observant Babylonian. But the two phrases cannot be
> regarded as having the same meaning; otherwise that
> Babylonian could have dispensed with his
> observations and contented himself with reflecting on
> the meanings of his words. The meanings, then,
> being different from one another, must be other than
> the named object, which is one and the same in both
> cases. (Quine, 1961, p. 9).

In the appropriate internal representation, there must be
two mental entities (concepts, nodes, or whatever)
corresponding to the two different intensions, morning star
and evening star. There is then an assertion about these
two intensional entities that they denote one and the same
external object (extension).

In artificial intelligence applications and psychology, it
is not sufficient for these intensions to be abstract entities
such as possibly infinite sets, but rather they must have
some finite representation inside the head as it were, or in
our case in the internal semantic representation.

F. Attributes and Values

Much of the structure of semantic networks is based on,

or at least similar to, the notion of attribute and value
which has become a standard concept in a variety of
computer science applications and which was the basis of
Raphael's SIR program (Raphael, 1964)--perhaps the earliest
forerunner of today's semantic networks.   Facts about an
object can frequently be stored on a "property list" of the
object   by   specifying   such   attribute-value   pairs   as
HEIGHT : 6 FEET, HAIRCOLOR : BROWN, OCCUPATION :
SCIENTIST, etc. (Such lists are provided, for example, for
all atoms in the LISP programming language.) One way of
thinking of these pairs is that the attribute name (i.e., the
first element of the pair) is the name of a "link" or
"pointer" which points to the "value" of the attribute (i.e.,
the second element of the pair).   Such a description of a
person named John might be laid out graphically as:

        JOHN
                HEIGHT   6 FEET
                HAIRCOLOR   BROWN
                OCCUPATION   SCIENTIST

    Now it may seem the case that the intuitive examples
which I just gave are all that it takes to explain what is
meant by the notion of attribute-value pair, and that the
use of such notations can now be used as part of a
semantic network notation without further explanation.   I
will try to make the case that this is not so and thereby
give a simple introduction to the kinds of things I mean
when I say that the semantics of the network notation need
to be specified.
    The above examples seem to imply that the thing which
occurs as the second element of an attribute-value pair is
the *name* or at least some unique handle on the value of
that attribute.   What will I do, however, with an input
sentence "John's height is greater than 6 feet?"   Most
people would not hesitate to construct a representation such
as:

        JOHN
            HEIGHT (GREATERTHAN 6 FEET)

Notice,   however,   that   our   interpretation   of   what   our

network notations mean has just taken a great leap. No
longer is the second element of the attribute-value pair a
name or a pointer to a value, but rather it is a predicate
which is asserted to be true of the value. One can think
of the names such as 6 FEET and BROWN in the previous
examples as special cases of identity predicates which are
abbreviated for the sake of conciseness, and thereby
consider the thing at the end of the pointer to be always a
predicate rather than a name. Thus there are at least two
possible interpretations of the meaning of the thing at the
end of the link--either as the name of the value or as a
predicate which must be true of the value. The former
will not handle the (GREATERTHAN 6 FEET) example,
while the latter will.

Let us consider now another example--"John's height is
greater than Sue's." We now have a new set of problems.
We can still think of a link named HEIGHT pointing from
JOHN to a predicate whose interpretation is "greater than
Sue's height", but what does the reference to Sue's height
inside this predicate have to do with the way that we
represented John's height? In a functional form we would
simply represent this as HEIGHT(JOHN) > HEIGHT(SUE),
or in LISP type "Cambridge Polish" notation,

(GREATER (HEIGHT JOHN)(HEIGHT SUE))

but that is departing completely from the notion of
attribute-value links. There is another possible
interpretation of the thing at the end of the HEIGHT link
which would be capable of dealing with this type of
situation. That is, the HEIGHT link can point from JOHN
to a node which represents the intensional object "John's
height". In a similar way, we can have a link named
HEIGHT from SUE to a node which represents "Sue's
height" and then we can establish a relation GREATER
between these two intensional nodes. (Notice that even if
the heights were the same, the two intensional objects
would be different, just as in the morning star/evening star
example.) This requires a major reinterpretation of the
semantics of our notation and a new set of conventions for
how we set up networks. We must now introduce a new
intensional node at the end of each attribute link and then

establish predicates as facts that are true about such intensional objects. It also raises for us a need to somewhere indicate about this new node that it was created to represent the concept of John's height, and that the additional information that it is greater than Sue's height is not one of its defining properties but rather a separate assertion about the node. Thus a distinction between defining and asserted properties of the node become important here. In my conception of semantic networks I have used the concept of an EGO link to indicate for the benefit of the human researcher and eventually for the benefit of the system itself what a given node is created to stand for. Thus the EGOs of these two nodes are John's height and Sue's height respectively. The EGO link represents the intensional identity of the node.

G. Links and Predication

In addition to considering what is at the end of a link, we must also consider what the link itself means. The examples above suggest that an attribute link named Z from node X to Y is equivalent to the English sentence "the Z of X is Y" or functionally $Z(X)=Y$ or (in the case where Y is a predicate) $Y(Z(X))$, (read Y of Z of X). Many people, however, have used the same mechanism and notation (and even called it attribute-value pairs) to represent arbitrary English verbs by storing a sentence such as "John hit Mary" as a link named HIT from the node for John to the node for Mary, as in the structure:

        JOHN
            HIT   MARY

and perhaps placing an inverse link under Mary:

        MARY
            HIT*   JOHN

If we do this, then suddenly the semantics of our notation has changed again. No longer do the link names stand for attributes of the node, but rather arbitrary relations

between the node and other nodes. If we are to mix the
two notations together as in:

```
JOHN
      HEIGHT        6 FEET
      HIT           MARY
```

then we need either to provide somewhere an indication
that these two links are of different types and therefore
must be treated differently by the procedures which make
inferences in the net, or else we need to find a unifying
interpretation such as considering that the "attribute"
HEIGHT is now really an abbreviation of the relation
"height of equals" which holds between JOHN and (the
node?) 6 FEET. It is not sufficient to leave it to the
intuition of the reader, we must know how the machine
will know to treat the two arcs correctly.

If we use Church's lambda notation, which provides a
convenient notation for naming predicates and functions
constructed out of combinations or variations of other
functions (this is used, for example, as the basic function
specification notation in the LISP programming language),
we could define the meaning of the height link as the
relation (LAMBDA $(X\ Y)$ (EQUAL (HEIGHT X) Y)). By
this we mean the predicate of two arguments X and Y
which is true when and only when the height of X is equal
to Y. Thus a possible unifying interpretation of the
notation is that the link is always the name of a relation
between the node being described and the node pointed to,
(providing that we reinterpret what we meant by the
original link named HEIGHT). Whatever we do, we clearly
need some mechanism for establishing relations between
nodes as facts (e.g., to establish the above GREATER
relation between the nodes for John's height and Sue's
height).

H. Relations of More Than Two Arguments

In the example just presented, we have used a link to
assert a relation between two objects in the network
corresponding to the proposition that John hit Mary. Such

a method of handling assertions has a number of
disadvantages, perhaps the simplest of which is that it is
constrained to handling binary relations. If we have a
predicate such as the English preposition "between" (i.e.,
(LAMBDA $(X\ Y\ Z)$ (Y is between X and Z))), then we
must invent some new kind of structure for expressing such
facts. A typical, but not very satisfying, notation which
one might find in a semantic network which uses links for
relations is something like:

Y

LOCATION   (BETWEEN X Z)

usually without further specification of the semantics of
the notation or what kind of thing the structure
(BETWEEN X Z) is. For example, is it the name of a
place? In some implementations it would be exactly that,
in spite of the fact that an underlying model in which
there is only one place between any given pair of places is
an inadequate model of the world we live in. Another
possible interpretation is that it denotes the range of places
between the two endpoints (this interpretation requires
another interpretation of what the LOCATION link means--
the thing at the end is no longer a name of a place but
rather a set of places, and the LOCATION link must be
considered to be implicitly existentially quantified in order
to be interpreted as asserting that the location is actually
one of those places and not all).

Given the notion which we introduced previously that
interprets the thing at the end of the link as a predicate
which must be true of the location, we have perhaps the
best interpretation--we can interpret the expression
(BETWEEN X Z) at the end of the link as being an
abbreviation for the predicate (LAMBDA (U) (BETWEEN X
U Z)), i.e., a one place predicate whose variable is U and
whose values of X and Z are fixed to whatever X and Z
are.

Although this representation of the three-place predicate
"between" (when supplied with an appropriate interpretation
of what it means) seems plausible, and I see no major
objections to it on the grounds of logical inadequacy, one is
left with the suspicion that there may be some predicates

of more than two places which do not have such an
intuitively satisfying decomposition into links connecting
only two objects at a time. For example, I had to
introduce the concept of location as the name of the link
from Y to the special object (BETWEEN X Z). In this
case, I was able to find a preexisting English concept which
made the creation of this link plausible, but is this always
the case? The account would have been much less
satisfying if all I could have produced was something like:

$$X$$
$$\text{BETWEEN1} \quad \text{(BETWEEN2 } Y \text{ } Z\text{)}$$

with an explication of its semantics that (BETWEEN2 Y Z)
was merely some special kind of entity which when linked
to X by a BETWEEN1 link represented the proposition
(BETWEEN X Y Z). It may be the case that all predicates
in English with more than two arguments have a natural
binary decomposition. The basic subject-predicate distinction which seems to be made by our language gives some
slight evidence for this. It seems to me, however, that
finding a natural binary decomposition for sentences such
as "John sold Mary a book" (or any of Schank's various
TRANS operations) is unlikely.

I. Case Representations in Semantic Networks

Another type of representation is becoming popular in
semantic networks and handles the problem of relations of
more than one argument very nicely. This representation is
based on the notion of case introduced by Fillmore (1968).
Fillmore advocates a unifying treatment of the inflected
cases of nominals in Latin and other highly inflected
languages and the prepositions and positional clues to role
that occur in English and other largely noninflected
languages. A *case* as Fillmore uses the term is the name of
a particular role that a noun phrase or other participant
takes in the state or activity expressed by the verb of a
sentence. In the case of the sentence "John sold Mary a
book" we can say that John is the *agent* of the action,
Mary is the *recipient* or *beneficiary* of the action, and the

book is the *object* or *patient* of the action (where I have
taken arbitrary but typical names for the case roles
involved for the sake of illustration). When such a
notation is applied to semantic network representations, a
major restructuring of the network and what it means to be
a link takes place. Instead of the assertion of a fact being
carried by a link between two nodes, the asserted fact is
itself a node. Our structure might look something like:

```
SELL
     AGT          JOHN
     RECIP        MARY
     PAT          BOOK
```

(ignoring for the moment what has happened to turn "a
book" into BOOK or for that matter what we mean by
JOHN and MARY--we will get into that later). The
notation as I have written it requires a great deal of
explanation, which is unfortunately not usually spelled out
in the presentation of a semantic network notation. In our
previous examples, the first item (holding the position
where we have placed SELL above) has been the unique
name or "handle" on a node, and the remaining link-value
pairs have been predicates that are true of this node. In
the case above, which I have written that way because one
is likely to find equivalent representations in the literature,
we are clearly not defining characteristics of the general
verb "sell", but rather setting up a description of a
particular instance of selling. Thus to be consistent with
our earlier format for representing a node we should more
properly represent it as something like:

```
S13472
     VERB         SELL
     AGT          JOHN
     RECIP        MARY
     PAT          BOOK
```

where S13472 is some unique internal handle on the node
representing this instance of selling, and SELL is now the
internal handle on the concept of selling. (I have gone
through this two-stage presentation in order to emphasize

that the relationship between the node S13472 and the concept of selling is not essentially different at this level from the relationship it has to the other nodes which fill the cases.)

J. Assertional and Structural Links

Clearly the case structure representation in a semantic network places a new interpretation on the nodes and arcs in the net. We still seem to have the same types of nodes that we had before for JOHN, MARY, etc., but we have a new type of node for nodes such as S13472 which represent assertions or facts. Moreover, the import of the links from this new type of node is different from that of our other links. Whereas the links which we discussed before are assertional, i.e., their mere presence in the network represents an assertion about the two nodes that they connect, these new link names, VERB, AGT, RECIP, PAT, are merely setting up parts of the proposition represented by node S13472, and no single link has any assertional import by itself; rather these links are definitional or structural in the sense that they constitute the definition of what node S13472 means.

Now you may argue that these links are really the same as the others, i.e., they correspond to the assertion that the agent of S13472 is JOHN and that S13472 is an instance of selling, etc. just like the "hit" link between John and Mary in our previous example. In our previous example, however, the nodes for John and for Mary had some a priori meanings independent of the assertion of hitting that we were trying to establish between them. In this case, S13472 has no meaning other than that which we establish by virtue of the structural links which it has to other nodes. That is, if we were to ask for the ego of the node S13472, we would get back something like "I am an instance of John selling a book to Mary" or "I am an instance of selling whose agent is John, whose recipient is Mary and whose patient is a book." If we were to ask for the ego of JOHN, we would get something like "I am the guy who works in the third office down the hall, whose name is John Smith, etc." The fact which I am trying to assert

with the "hit" link is not part of the ego of JOHN or else
I would not be making a new assertion.

This difference between assertional and structural links
is rather difficult for some people to understand, and is
often confused in various semantic network representations.
It is part of the problem that we cited earlier in trying to
determine whether a structure such as:

```
N12368
      SUPERC          TELEPHONE
      MOD             BLACK
```

is to be interpreted as an intensional representation of a
black telephone or an assertion that telephones are black.
If it is to be interpreted as an intensional representation of
the concept of a black telephone, then both of these links
are structural or definitional. If on the other hand, it is
to be interpreted as asserting that telephones are black,
then the first link is structural while the second is
assertional. (The distinction between structural and
assertional links does not take care of this example entirely
since we still have to worry about how the assertional link
gets its quantificational import for this interpretation, but
we will discuss this problem later.)

The above discussion barely suffices to introduce the
distinction between structural and assertional links, and
certainly does not make the distinction totally clear.
Moreover, before we are through, we may have cause to
repudiate the assumption that the links involved in our
non-case representation should be considered to have
assertional import. Perhaps the best way to get deeper into
the problems of different types of links with different
imports and the representation of intensional entities is to
consider further some specific problems in knowledge
representation.

IV. PROBLEMS IN KNOWLEDGE REPRESENTATION

In previous sections I hope that I have made the point
that the same semantic network notations could be used by

different people (or even by the same person at different
times for different examples) to mean different things, and
therefore one must be specific in presenting a semantic
network notation to make clear what one means by the
notations which one uses (i.e., the semantics of the
notation). In the remainder of this chapter, I would like
to discuss two difficult problems of knowledge
representation and use the discussion to illustrate several
additional possible uses of links and some of the different
types of nodes and links which are required in a semantic
network if it is to serve as a medium for representing
human verbal knowledge. The specific problems which I
will consider are the representation of restrictive relative
clauses and the representation of quantified information.

A. Relative Clauses

In attaching modifiers to nodes in a network to provide
an intensional description for a restricted class, one often
requires restrictions which do not happen to exist in the
language as single-word modifiers but have to be
constructed out of more primitive elements. The relative
clause mechanism permits this. Anything that can be said
as a proposition can be used as a relative clause by leaving
some one of its argument slots unfilled and using it as a
modifier. (We will be concerned here only with restrictive
relative clauses and not those which are just parenthetical
comments about an already determined object.) Let me
begin my discussion of relative clauses by dispensing with
one inadequate treatment.

The Shared Subpart Fallacy: A mechanism which
occasionally surfaces as a claimed technique for dealing
with relative clauses is to take simply the two propositions
involved, the main clause and the relative clause, and
represent the two separately as if they were independent
propositions. In such a representation, the sentence "The
dog that bit the man had rabies" would look something like
that in Fig. 1. The point of interest here is not the names
of the links (for which I make no claims) nor the type of
representation (case oriented, deep conceptual, or whatever),

but simply the fact that the only relationship between the two propositions is that they share the same node for dog. There are a number of problems with this representation: First, since there is no other relationship between the two sentences except sharing of a node (which is a symmetric relationship) there is no indication of which is the main clause and which is the relative clause. That is, we would get the same internal representation for the sentence "The dog that had rabies bit the man."

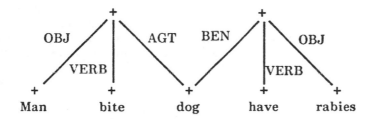

Fig. 1. A shared subpart representation.

Another difficulty is that there is nothing to indicate that the two sentences go together at all in a relative clause relationship. It is possible that on two different occasions we were told about this dog. On one occasion that he had rabies and on another that he bit a man. Then the presence of the two propositions in our data base both sharing the same node for dog would give us a structure identical to that for the example sentence. Now there is a subtle confusion which can happen at this point which I would like to try to clarify. You may say to me, "So what is the problem? Suppose I tell you about this dog and suppose I have told you the two facts at different times, then it is still true that the dog that bit the man has rabies." How do I answer such an argument? On the face of it it seems true. Yet I maintain that the argument is fallacious and that it results from too shallow a treatment of the issues. The crux of the matter I think rests in the notion of which dog we are talking about. Unfortunately, this issue is one that gets omitted from almost all such discussions of semantic networks. If the two facts were told to me at different times, how did I know that they

were about the same dog? (Without further explication of
the semantics of the network notation, it is not even clear
that we are talking about a particular dog and not about
dogs in general.) It is exactly in order to relate the second
fact to the first that we need the relative clause
mechanism. In the next section we will consider the
problem in more detail.

The Transient-Process Account: Quillian[1] once made the
observation that a portion of what was in an input sentence
was essentially stage directions used to enable the
understanding process to identify an appropriate internal
concept or node and the rest of the utterance was to be
interpreted as new information to be added somehow to the
network (and similar observations have been made by
others). This gives an attractive account of the relative
clause problem above. We interpret the relative clause not
as something to be added to the network at all, but rather
as a description to be used by the understander to
determine which dog is in question. After this, we can
forget about the relative clause (it has served its
usefulness) and simply add the new information to the
network. We might call this the "transient-process
account". Under this account, if I was told about a dog
that bit a man and later told that the dog that bit the
man had rabies, then I would simply use the relative clause
to find the internal concept for the dog that bit the man,
and then add the new information that the dog had rabies.
What's wrong with that account? Doesn't that explain
everything?
Well, no. First, it simply evades the issue of
representing the meaning of the sentence, focusing instead
on the resulting change in memory contents. It says
essentially that the role of the relative clause is a
temporary and transient one that exists only during the
processing of the utterance and then goes away. But you
say, "well, isn't that a plausible account, does not that take
care of the problem nicely, who says you have to have a
representation of the original sentence anyway?"

---

[1]Personal communication.

Let's start from the first question--yes, it is a plausible account of the interpretation of *many* sentences, including this one in the context I just set up, and it may also be a correct description of what happens when humans process such sentences. It does not, however, take care of all occurrences of relative clauses. What about a situation when I read this sentence out of context and I haven't heard about the dog before? Then my processing must be different. I must infer that there must be a dog that I do not know about, perhaps create a new node for it, and then assert about this new node that it has rabies. Clearly also I must associate with this new node that it is a dog and that it bit a man. How then do I keep these two different types of information separate--the information which designates what I set the node up to stand for and that which the sentence asserted about it. We're back to the same problem. We need to distinguish the information that is in the relative clause from that in the main clause.

One possible way would be the use of an EGO link which points to a specification of what the node represents. Using such a link, when one creates the new node for the dog which bit the man, one would give the new node an EGO link which in essence says "I am the node which represents the dog that bit the man." When one then adds information to this node asserting additional facts about it, the original motivation for creating the node in the first place is not forgotten and the difference between the sentences "The dog that bit the man had rabies" and "The dog that had rabies bit the man" would lie in whether the facts about biting or about rabies were at the end of the EGO link. (There are a number of other questions which would require answers in order to complete the specification of the use of EGO links for this purpose--such as whether the propositions at the end of the EGO link are thereby made indirectly available as properties of this node or whether they are redundantly also included in the same status as the additional asserted properties which come later. We will not, however, go into these issues here.)

The above argument should have convinced you that the simple explanation of using relative clauses always only to identify preexisting nodes does not cover all of the cases. For certain sentences such as the above example, the object

determined by the relative clause does not previously exist
and something must be created in the semantic network
which will continue to exist after the process is finished.
This thing must have an internal representation which
preserves the information that it is an object determined by
a relative clause.

A second argument against the transient process account
is that even for sentences where nothing needs to remain in
memory after the process has completed (because the
relative clause has been used to locate a preexisting node),
something needs to be extracted from the input sentence
which describes the node to be searched for. In our
previous example something like the proposition "the dog
bit the man" needs to be constructed in order to search for
its instances, and the process must know when it finds such
an instance that it is the dog that is of interest and not
the man. This specification of the node to be searched for
is exactly the kind of thing which a semantic interpretation
for the noun phrase "the dog that bit the man" should be.
Thus even when no permanent representation of the relative
clause needs to remain after the understanding process has
completed, something equivalent to it still needs to be
constructed as part of the input to the search process. The
transient process account does not eliminate the need for
such a representation, and the issue of whether a complete
representation of the entire sentence (including the relative
clause) gets constructed and sent off to the understanding
process as a unit or whether small pieces get created and
sent off independently without ever being assembled into a
complete representation is at this point a red herring. The
necessary operations which are required for the search
specification are sufficient to construct such a
representation, and whether it is actually constructed or
whether parts of it are merely executed for effect and then
cast away is a totally separate question.

A third argument against the transient process account,
which should have become apparent in the above discussion,
is that it is not an account at all, but merely a way of
avoiding the problem. By claiming that the relative clause
is handled during the transient process we have merely
pushed the problem of accounting for relative clauses off
onto the person who attempts to characterize the

understanding process. We have not accounted for it or solved it.

B. Representation of Complex Sentences

Let us return to the question of whether one needs a representation of the entire sentence as a whole or not. More specifically, does one need a representation of a proposition expressed about a node which itself has a propositional restriction, or can one effectively break this process up in such a way that propositions are always expressed about definite nodes? This is going to be a difficult question to answer because there is a sense in which even if the answer is the former, one can model it with a process which first constructs the relative clause restricted node and then calls it definite and represents the higher proposition with a pointer to this new node. The real question, then, is in what sense is this new node definite? Does it always refer to a single specific node like the dog in our above example, or is it more complicated than that? I will argue the latter.

C. Definite and Indefinite Entities

Consider the case which we hypothesized in which we had to infer the existence of a heretofore unknown dog because we found no referent for "the dog that bit the man". This new node still has a certain definiteness to it. We can later refer to it again and add additional information, eventually fleshing it out to include its name, who owns it, etc. As such it is no different from any other node in the data base standing for a person, place, thing, etc. It got created when we first encountered the object denoted (or at least when we first recognized it and added it to our memory) and has subsequently gained additional information and may in the future gain additional information still. We know that it is a particular dog and not a class of dogs and many other things about it.

Consider, however, the question "Was the man bitten by

a dog that had rabies?" Now we have a description of an indefinite dog and moreover we have not asserted that it exists but merely questioned its existence. Now you may first try to weasel out of the problem by saying something like, "Well, what happens is that we look in our data base for dogs that have rabies in the same way that we would in the earlier examples, and finding no such dog, we answer the question in the negative." This is another example of pushing the problem off onto the understanding process; it does not solve it or account for it, it just avoids it (not to mention the asumption that the absence of information from the network implies its falsity).

Let us consider the process more closely. Unless our process were appropriately constructed (how?) it would not know the difference (at the time it was searching for the referent of the phrase) between this case and the case of an assertion about an unknown dog. Hence the process we described above would create a new node for a dog that has rabies unless we block it somehow. Merely asking whether the main clause is a question would not do it, since the sentence "Did the dog that bit the man have rabies?" still must have the effect of creating a new definite node. (This is due to the effect of the presupposition of the definite singular determiner "the" that the object described must exist.) Nor is it really quite the effect of the indefinite article "a", since the sentence "a dog that had rabies bit the man" should still create a definite node for the dog. We could try conditions on questioned indefinites. Maybe that would work, but let me suggest that perhaps you do not want to block the creation of the new node at all but rather simply allow it to be a different type of entity, one whose existence in the real world is not presupposed by an intensional existence in the internal semantic network.

If we are to take this account of the hypothetical dog in our question, then we have made a major extension in our notion of structures in a semantic network and what they mean. Whereas previously we construed our nodes to correspond to real existing objects, now we have introduced a new type of node which does not have this assumption. Either we now have two very different types of nodes (in which case we must have some explicit indicator or other mechanism in the notation to indicate the type of every node) or else we must impose a unifying interpretation. If we have two different types of nodes, then we still have the problem of telling the process which constructs the nodes which type of node to construct in our two examples.

One possible unifying interpretation is to interpret every node as an intensional description and assert an explicit predicate of existence for those nodes which are intended to correspond to real objects. In this case, we could either rely on an implicit assumption that intensional objects used as subjects of definite asserted sentences (such as "the dog that bit the man had rabies") must actually exist, or we could postulate an inferential process which draws such inferences and explicitly asserts existence for such entities.

Since the above account of the indefinite relative clause in our example requires such a major reinterpretation of the fundamental semantics of our network notations, one might be inclined to look for some other account that was less drastic. I will argue, however, that such internal intensional entities are required in any case to deal with other problems in semantic representation. For example, whenever a new definite node gets created, it may in fact stand for the same object as some other node which already exists, but the necessary information to establish the identity may only come later or not at all. This is a fundamental characteristic of the information that we must store in our nets. Consider again Frege's morning star / evening star example. Even such definite descriptions, then, are essentially intensional objects. (Notice as a consequence that one cannot make negative identity assertions simply on the basis of distinctness of internal semantic representations.)

Perhaps the strongest case for intensional nodes in semantic networks comes from verbs such as "need" and "want". When one asserts a sentence such as "I need a wrench", one does not thereby assert the existence of the object desired. One must, however, include in the representation of this sentence some representation of the thing needed. For this interpretation, the object of the verb "need" should be an intensional description of the needed item. (It is also possible for the slot filler to be a node designating a particular entity rather than just a description, thus giving rise to an ambiguity of interpretation of the sentence. That is, is it a particular wrench that is needed, or will any wrench do?)

D. Consequences of Intensional Nodes

We conclude that there must be some nodes in the semantic network which correspond to descriptions of entities rather than entities themselves. Does that fix up the problem? Well, we have to do more than just make the assumption. We have to decide how to tell the two kinds of nodes apart, how we decide for particular sentences which type to create, and how to perform inferences on these nodes. If we have nodes which are intensional descriptions of entities, what does it mean to associate properties with the nodes or to assert facts about the nodes. We cannot just rely on the arguments that we made when we were assuming that all of the nodes corresponded to definite external entities. We must see whether earlier interpretations of the meanings of links between nodes still hold true for this new expanded notion of node or whether they need modification or reinterpretation. In short we must start all over again from the beginning but this time with attention to the ability to deal with intensional descriptions.

Let me clarify further some of the kinds of things which we must be able to represent. Consider the sentence "Every boy loves his dog". Here we have an indefinite node for the dog involved which will not hold still. Linguistically it is marked definite (i.e., the dog that belongs to the boy), but it is a variable definite object whose reference changes with the boy. There are also variable entities which are indefinite as in "Every boy needs a dog." Here we plunge into the really difficult and crucial problems in representing quantification. It is easy to create simple network structures that model the logical syllogisms by creating links from subsets to supersets, but the critical cases are those like the above. We need the notion of an intensional description for a variable entity.

To summarize, then, in designing a network to handle intensional entities, we need to provide for definite entities that are intended to correspond to particular entities in the real world, indefinite entities which do not necessarily have corresponding entities in the real world, and definite and indefinite variable entities which stand in some relation to other entities and whose instantiations will depend on the instantiations of those other entities.

E. Functions and Predicates

Another question about the interpretation of links and what we mean by them comes in the representation of information about functions and predicates. Functions and predicates have a characteristic that clearly sets them apart from the other types of entities which we have mentioned (with the possible exception of the variable entity which depends on others)--namely, they take arguments. Somewhere in the internal representation of an entity which is a function or a predicate there must be information about the arguments which the function or predicate takes, what kinds of entities can fill those arguments, and how the value of the function or the truth of the predicate is determined or constrained by the values of the arguments. There is a difference between representing the possible entities that can serve as arguments for a predicate and expressing the assertion of the predicate for particular values or classes of values of those arguments. Unfortunately this distinction is often confused in talking about semantic networks. That is, it is all too easy to use the notation:

```
LOVE
        AGT          HUMAN
        RECIP        HUMAN
```

to express constraints on the possible fillers for the arguments of the predicate and to use the same link names in a notation such as:

```
S76543
        VERB         LOVE
        AGT          JOHN
        RECIP        SALLY
```

to represent the assertion that John loves Sally. Here we
have a situation of the same link names meaning different
things depending on the nodes which they are connected to.

Without some explicit indication in the network
notation that the two nodes are of different types, no
mechanical procedure operating on such a network would be
able to handle these links correctly in both cases. With an
explicit indication of node type and an explicit definition
that the meaning of an arc depends on the type of the
node to which it is connected (and how), such a procedure
could be defined, but a network notation of this sort would
probably be confusing as an explanatory device for human
consumption. This is functionally equivalent, however, to
an alternative mechanism using a dual set of links with
different names (such as R-AGT and AGT, for example)
which would make the difference explicit to a human
reader and would save the mechanical procedure from
having to consult the type of the node to determine the
import of the link. Notice that in either case we are
required to make another extension of the semantics of
our network notation since we have two different kinds of
links with different kinds of import. The ones which make
statements about possible slot fillers have assertional
import (asserting facts about the predicate LOVE in this
case) while the ones that make up the arguments of S76543
have structural import (building up the parts of the
proposition, which incidentally may itself not be asserted
but only part of some intensional representation).

We conclude that the difference between the
specification of possible slot fillers for a predicate as part
of the information about the predicate and the specification
of particular slot fillers for particular instances of the
predicate requires some basic distinction in our semantic
network notation. One is left with several questions as to
just how this distinction is best realized (for example does
one want a dual set of link names--or is there a preferable
notation?). For the moment, however, let us leave those
questions unexplored along with many issues that we have
not begun to face and proceed with another problem of
knowledge representation that imposes new demands on the
interpretations of links and the conventions for
representing facts in semantic networks.

F. Representing Quantified Expressions

The problem of representing quantified information in semantic networks is one that few people have faced and even fewer handled adequately. Let me begin by laying to rest a logically inadequate way of representing quantified expressions which unfortunately is the one most used in implemented semantic networks. It consists of simply tagging the quantifier onto the noun phrase it modifies just as if it were an adjective. In such a notation, the representation for "every integer is greater than some integer" would look something like:

```
S11113
        VERB          GREATER
        ARG1          D12345
        ARG2          D67890

D12345
        NOUN          INTEGER
        MOD           EVERY

D67890
        NOUN          INTEGER
        MOD           SOME
```

Now there are two possible interpretations of this sentence depending on whether or not the second existential quantifier is considered to be in the scope of the universal quantifier. In the normal interpretation, the second integer depends on the first and the sentence is true, while a pathological interpretation of the sentence is that there is some integer which every integer is greater than. (Lest you divert the issue with some claim that there is only one possible interpretation taking the quantifiers in the order in which they occur in the sentence consider a sentence such as "Everybody jumped in some old car that had the keys in it", in which the normal interpretation is the opposite.) Since our semantic network notation must provide a representation for whichever interpretation we decide was meant, there must be some way to distinguish the difference. If anything, the representation we have given seems to suggest the interpretation in which there is some integer that every integer is greater than. If we take this as the interpretation of the above notation, then we need another representation for the other (and in this case correct) interpretation--the one in which the second integer is a variable entity dependent on the first.

To complicate matters even further, consider the case of numerical quantifiers and a sentence such as "three lookouts saw two boats". There are three possible interpretations of the quantifiers in this case. In the one that seems to correspond to treating the quantifier as a modifier of the noun phrase, we would have one group of three lookouts that jointly participated in an activity of seeing one group of two boats. There is, however, another interpretation in which each of three lookouts saw two boats (for an unknown total number of boats between 2 and 6 since we are not told whether any of them saw the same boats as the others) and still another interpretation in which each of two boats was seen by three men. We must have a way in our network notation to represent unambiguously all three of these possible interpretations. Quillian's (1968) suggestion of using "criterialities" on the arcs to indicate quantification will fail for the same reasons unless some mechanism for indicating which arguments depend on which others is inserted.

Before proceeding to discuss logically adequate ways of dealing with quantification, let me also lay to rest a borderline case. One might decide to represent the interpretation of the sentence in which each of three men saw two boats, for example, by creating three separate nodes for the men and asserting about each of them that he saw two boats. This could become logically adequate if the appropriate information were indicated that the three men were all different (it is not adequate to assume that internal nodes are different just because they are different nodes--recall the morning star/evening star example) and if the three separate facts are tied together into a single fact somewhere (e.g., by a conjunction) since otherwise this would not be an expression of a single fact (which could be denied, for example). This is, however, clearly not a reasonable representation for a sentence such as "250 million people live in the United States", and would be a logical impossibility for representing universally quantified expressions over sets whose cardinality was not known.

A variant of this is related to the transient process account. One might argue that it is not necessary to represent a sentence such as "Every boy has a dog" as a unit, but one can simply add an assertion to each internal node representing a boy. To be correct, however, such an account would require a network to have perfect knowledge (i.e., an internal node for every boy that exists in the world), a practical impossibility. We cannot assume that the entities in our network exhaust those that exist in the world. Hence we must represent this assertion in a way that will apply to future boys that we may learn about and not just to those we know about at this moment. To do this we must be able to store an intensional representation of the universally quantified proposition.

Quantifiers as Higher Operators: The traditional representation of quantifiers in the predicate calculus is that they are attached to the proposition which they govern in a string whose order determines the dependency of the individual variables on other variables. Thus the two interpretations of our first sentence are:

$(\forall X/\text{integer})$ $(\exists Y/\text{integer})$ $(\text{GREATER } X \ Y)$
and
$(\exists Y/\text{integer})$ $(\forall X/\text{integer})$ $(\text{GREATER } X \ Y)$

where I have chosen to indicate explicitly in the quantifier prefix the range of quantification of the variable (see Woods, 1967) for a discussion of the advantages of doing this--namely the uniform behavior for both universal and existential quantifiers). In the question-answering systems that I have constructed, including the LUNAR system, I have used a slightly expanded form of such quantifiers which uniformly handles numerical quantifiers and definite determiners as well as the classical universal and existential quantifiers. This formulation treats the quantifiers as higher predicates which take as arguments a variable name, a specification of the range of quantification, a possible restriction on the range, and the proposition to be quantified (which includes a free occurrence of the variable

of quantification and which may be already quantified by
other quantifiers). In this notation, the above two
interpretations would be represented as:

(FOR EVERY *X* / INTEGER : T ;
   (FOR SOME *Y* / INTEGER : T ; (GREATER *X Y*)))

and

(FOR SOME *Y* / INTEGER : T ;
   (FOR EVERY *X* / INTEGER : T ; (GREATER *X Y*)))

where the component of the notation following the ":" in
these expressions is a proposition which restricts the range
of quantification (in this case the vacuously true
proposition T) and the component following the ";" is the
proposition being quantified. This type of higher-operator
representation of quantification can be represented in a
network structure by creating a special type of node for the
quantifier and some special links for its components. Thus
we could have something like:

S39732		
	TYPE	QUANT
	QUANT-TYPE	EVERY
	VARIABLE	*X*
	CLASS	INTEGER
	RESTRICTION	T
	PROP	S39733
S39733		
	TYPE	QUANT
	QUANT-TYPE	SOME
	VARIABLE	*Y*
	RESTRICTION	T
	PROP	S39734
S39734		
	TYPE	PROPOSITION
	VERB	GREATER
	ARG1	*X*
	ARG2	*Y*

This is essentially the technique used by Shapiro (1971), who is one of the two people I know of to suggest a logically adequate treatment of quantifiers in his nets. (The other one is Martin Kay, whose proposal we will discuss shortly.) This technique has an unpleasant effect, however, in that it breaks up the chains of connections from node to node that one finds attractive in the more customary semantic network notations. That is, if we consider our sentence about lookouts and boats, we have gone successively from a simple-minded representation in which we might have a link labeled "see" which points from a node for "lookout" to one for "boats", to a case representation notation in which the chain becomes an inverse agent link from "lookout" to a special node which has a verb link to "see" and a patient link to "boats", and finally to a quantified representation in which the chain stretches from "lookout" via an inverse CLASS link to a quantifier node which has a PROP link to another quantifier node which has a CLASS link to "boats" and a PROP link to a proposition which has a VERB link to "see". Thus our successive changes in the network conventions designed to provide them with a logically adequate interpretation are carrying with them a cost in the directness of the associative paths. This may be an inevitable consequence of making the networks adequate for storing knowledge in general, and it may be that it is not too disruptive of the associative processing that one would like to apply to the memory representation. On the other hand it may lead to the conclusion that one cannot accomplish an appropriate associative linking of information as a direct consequence of the notation in which it is stored and that some separate indexing mechanism is required.

Other Possible Representations: There are two other possible candidates for representing quantified information, one of which to my knowledge has not been tried before in semantic networks. I will call them the "Skolem function method" and the "lambda abstraction method", after well-known techniques in formal logic.

Skolem Functions: The use of Skolem functions to represent quantified expressions is little known outside the field of mechanical theorem proving and certain branches of formal logic, but it is a pivotal technique in resolution theorem proving and is rather drastically different from the customary way of dealing with quantifiers in logic. The technique begins with a quantified expression containing no negative operators in the quantifier prefix (any such can be removed by means of the transformations exchanging "not every" for "some not" and "not some" for "every not"). It then replaces each instance of an existentially quantified variable with a functional designator whose function is a unique function name chosen for that existential variable and whose arguments are the universally quantified variables in whose scopes the existential quantifier for that variable lies. After this the existential quantifiers are deleted and, since the only remaining variables are universally quantified, the universal quantifiers can be deleted and free variables treated as implicitly universally quantified. The expression $(\forall x)(\exists y)(\forall z)(\exists w) \, P(x,y,z,w)$, for example, becomes $P(x,f(x),z,g(x,z))$, where f and g are new function names created to replace the variables y and w.

Notice that the arguments of the functions f and g in the result preserve the information about the universally quantified variables on which they depend. This is all the information necessary to reconstruct the original expression and is intuitively exactly that information which we are interested in to characterize the difference between alternative interpretations of a sentence corresponding to different quantifier orderings--i.e., does the choice of a given object depend on the choice of a universally quantified object or not? Thus the Skolem function serves as a device for recording the dependencies of an existentially quantified variable. An additional motivating factor for using Skolem functions to represent natural language quantification is that the quantification operation implicitly determines a real function of exactly this sort, and there are places in natural language dialogs where this implicit function appears to be referenced by anaphoric pronouns outside the scope of the original quantifier (e.g., in "Is there someone here from Virginia? If so, I have a prize for him", the "him" seems to refer to the value of such a function). We can obtain a semantic network notation based on this Skolem function analogy by simply including with every existentially quantified object a link which points to all of the universally quantified objects on which this one depends. This is essentially the technique proposed by Kay (1973).

It must be pointed out that one difficulty with the Skolem function notation which accounts for its little use as a logical representation outside the theorem proving circles is that it is not possible to obtain the negation of a Skolem form expression by simply attaching a negation operator to the "top". Rather, negation involves a complex operation which changes all of the universal variables to existential ones and vice versa and can hardly be accomplished short of converting the expression back to quantifier prefix form, rippling the negation through the quantifier prefix to the embedded predicate and then reconverting to Skolem form. This makes it difficult, for example, to store the denial of an existing proposition. It seems likely that the same technique of explicitly linking the quantified object to those other objects on which it depends might also handle the case of numerically quantified expressions although I am not quite sure how it would all work out--especially with negations.

Lambda Abstraction: We have already introduced Church's lambda notation as a convenient device for expressing a predicate defined by a combination or a modification of other predicates. In general, for any completely instantiated complex assemblage of predicates and propositions, one can make a predicate of it by replacing some of its specific arguments with variable names and embedding it in a lambda expression with those variables indicated as arguments. For example, from a sentence "John told Mary to get something and hit Sam" we can construct a predicate (LAMBDA (X) John told Mary to get something and hit X) which is true of Sam if the original sentence is true and may be true of other individuals as well. This process is called "lambda abstraction".

Now one way to view a universally quantified sentence such as "all men are mortal" is simply as a statement of a relation between a set (all men) and a predicate (mortal) --

namely that the predicate is true of each member of the set (call this relation FORALL). By means of lambda abstraction we can create a predicate of exactly the type we need to view *every* instance of universal quantification as exactly this kind of assertion about a set and a predicate. For example, we can represent our assertion that every integer is greater than some integer as an assertion of the FORALL relation between the set of integers and the predicate

(LAMBDA (X) (X is greater than some integer))

and in a similar way we can define a relation FORSOME which holds between a set and a predicate if the predicate is true for some member of the set, thus giving us a representation:

(FORALL INTEGER (LAMBDA (X)
    (FORSOME INTEGER (LAMBDA (Y)
    (GREATER X Y)))))

which can be seen as almost a notational variant of the higher-operator quantifier representation. Notice that the expression (LAMBDA (Y) (GREATER X Y)) is a predicate whose argument is Y and which has a free variable X. This means that the predicate itself is a variable entity which depends on X--i.e., for each value of X we get a different predicate to be applied to the Ys.

The use of this technique in a semantic network notation would require a special type of node for a predicate defined by the lambda operator, but such a type of node is probably required anyway for independent reasons (since the operation of lambda abstraction is an intellectual operation which one can perform and since our semantic network should be able to store the results of such mental gymnastics). The structure of the above expression might look like:

```
S12233
        TYPE            PROPOSITION
        VERB            FORALL
        CLASS           INTEGER
        PRED            P12234
```

```
P12234
        TYPE                PREDICATE
        ARGUMENTS           (X)
        BODY                S12235


S12235
        TYPE                PROPOSITION
        VERB                FORSOME
        CLASS               INTEGER
        PRED                P12236

P12236
        TYPE                PREDICATE
        ARGUMENTS           (Y)
        BODY                S12237

S12237
        TYPE                PROPOSITION
        VERB                GREATER
        ARG1                X
        ARG2                Y
```

V. CONCLUSION

In the preceding sections, I hope that I have illustrated by example the kinds of explicit understanding of what one intends various network notations to mean that must be made in order to even begin to ask the questions whether a notation is an adequate one for representing knowledge in general (although for reasons of space I have been more brief in such explanations in this chapter than I feel one should be in presenting a proposed complete semantic network notation). Moreover, I hope that I have made the point that when one does extract a clear understanding of the semantics of the notation, most of the existing semantic network notations are found wanting in some major respects--notably the representation of propositions without committment to asserting their truth and in representing various types of intensional descriptions of objects without commitment to their external existence, their external

distinctness, or their completeness in covering all of the
objects which are presumed to exist in the world. I have
also pointed out the logical inadequacies of almost all
current network notations for representing quantified
information and some of the disadvantages of some logically
adequate techniques.

I have not begun to address all of the problems that
need to be addressed, and I have only begun to discuss the
problems of relative clauses and quantificational
information. I have not even mentioned other problems
such as the representation of mass terms, adverbial
modification, probabilistic information, degrees of certainty,
time, and tense, and a host of other difficult problems.
All of these issues need to be addressed and solutions
integrated into a consistent whole in order to produce a
logically adequate semantic network formalism. No existing
semantic network comes close to this goal.

I hope that by focusing on the logical inadequacies of
many of the current (naive) assumptions about what
semantic networks do and can do, I will have stimulated
the search for better solutions and flagged some of the
false assumptions about adequacies of techniques that might
otherwise have gone unchallenged. As I said earlier, I
believe that work in the area of knowledge representation
in general, and semantic networks in particular, is
important to the further development of our understanding
of human and artificial intelligence and that many
essentially correct facts about human performance and
useful techniques for artificial systems are emerging from
this study. My hope for this chapter is that it will
stimulate this area of study to develop in a productive
direction.

REFERENCES

Carbonell, J. R., & Collins, A. M. Natural semantics in artificial intelligence. *Proceedings of Third International Joint Conference on Artificial Intelligence*, 1973, 344-351. (Reprinted in the *American Journal of Computational Linguistics*, 1974, *1*, Mfc. 3.)

Davis, M. *Computability and unsolvability*. New York: McGraw-Hill, 1958.

Fillmore, C. The case for case. In Bach and Harms (Eds.), *Universals in linguistic theory*, Chicago, Ill.: Holt, 1968.

Kay, M. The MIND system. In R. Rustin (Ed.) *Natural language processing*. New York: Algorithmics Press, 1973.

Lindsay, R. K. Inferential memory as the basis of machines which understand natural language. In E. A. Feigenbaum & J. Feldman (Eds.), *Computers and thought*. New York: McGraw-Hill, 1963.

Quillian, M. R. Semantic memory. In M. Minsky (Ed.), *Semantic information processing*. Cambridge, Mass.: MIT Press, 1968.

Quillian, M. R. The teachable language comprehender. *Communications of the Association for Computing Machinery*, 1969, *12*, 459-475.

Quine, W. V. *From a logical point of view*. (2nd Ed., rev.) New York: Harper, 1961.

Raphael, B. A computer program which 'understands'. *AFIPS Conference Proceedings*, 1964, *26*, 577-589.

Rumelhart, D. E., Lindsay, P. H., & Norman, D. A. A process model for long-term memory. In E. Tulving & W. Donaldson (Eds.), *Organization of memory*. New York: Academic Press, 1972.

Schank, R. C. *Conceptual information processing*. Amsterdam: North-Holland, 1975.

Shapiro, S. C. A net structure for semantic information storage, deduction, and retrieval. *Proceedings of the Second International Joint Conference on Artificial Intelligence*, 1971, 512-523.

Simmons, R. F. Semantic networks: Their computation and use for understanding English sentences. In R. C. Schank & K. M. Colby (Eds.), *Computer models of thought and language*. San Francisco, Ca.: Freeman, 1973.

Winograd, T. *Understanding natural language*. New York: Academic Press, 1972.

Woods, W.A. Semantics for a Question-Answering System. Ph.D. Thesis, Division of Engineering and Applied Physics, Harvard University, 1967. (Also in Report NSF-19, Harvard Computation Laboratory, September 1967. Available from NTIS as PB-176-548.)

Woods, W.A. Meaning and machines. *Proceedings of the International Conference on Computational Linguistics*, Pisa, Italy, August 1973(a).

Woods, W.A. Progress in natural language understanding: An application to lunar geology. *AFIPS Conference Proceedings*, 1973(b), *42*, 441-450.

Woods, W.A. Syntax, semantics, and speech, presented at IEEE Symposium on Speech recognition, Carnegie-Mellon University, April 1974.

Woods, W. A., Kaplan, R. M., & Nash-Webber, B. The lunar sciences natural language information system: final report. BBN Report No. 2378, June 1972.

REFLECTIONS ON THE FORMAL DESCRIPTION
OF BEHAVIOR

Joseph D. Becker
Bolt Beranek and Newman
Cambridge, Massachusetts

I. INTRODUCTION

Mathematics arises from man's attempt to find concepts that allow a wide range of phenomena to be described in similar terms, and thereby understood in a coherent manner. For example, primitive man found that the basic concept of *number* could tell him something about days and kinsmen and bowls of food. More recently, the concepts of *accretion* and *rate* have given us the calculus, which allows us to see a flywheel, a water tank, and a capacitor as embodiments of a single principle. In the past two decades, the field of computer programming has given new and precise meanings to many concepts that formerly were unrelated items in the general vocabulary: *interrupt, scheduling, backtracking, hierarchical organization,* and so forth. Besides serving as useful tools for the computer programmer, these concepts are gradually being explored by psychologists as a means of describing the behavior of men and beasts.

The purpose of this chapter is to elucidate and interrelate some of these concepts from the computer realm that are useful in organizing our understanding of human and animal behavior. Our point of view will generally be that of a psychologist trying to describe observed animate behavior, but we will use the neutral term "behavioral system" to indicate that these concepts can be applied to animate and inanimate systems alike.

The tone of this chapter is by no means formal, mathematical, or rigorous. Our intention is merely to set forth some concepts and explore their origins, their ramifications, and their interconnections. Hopefully, this exploration may provoke an insight or two in the reader, even if he or she is already familiar with the concepts themselves.

II. HIERARCHICAL ORGANIZATION OF PROCESSES

Any behavior that we observe must unfold linearly with time. Yet often we describe or design a behavioral system in terms of a hierarchy of processes. Why do we not represent every system simply as a linear sequence of actions? The reason is that we are able to see significant recurring patterns in a linear sequence of events, and we attribute the appearance of similar subsequences to the presence of a single "subprocess". That is, we form the concepts of individual subprocesses, such as "squaring a number" or "grasping an object", by *induction over time*, in precisely the same manner that we form object-concepts such as "cat" or "sunset". Process-concepts naturally nest to form part-whole hierarchies, e.g., "grasping an object" is part of "picking up an object", just as "pink clouds" are a part of "sunset".

Because of our experience with hierarchically *structured* systems (e.g., computer programs and human management structures), we tend to think of hierarchical *behavioral* organization as being similarly "real", i.e., part of the mechanism that actually generates the behavior. This need not necessarily be the case. We can in fact take any activity, such as "grasping an object", and break it down further into "opening the hand", "orienting the hand", "moving the hand to the object", etc. This analysis does not mean that grasping *actually* proceeds in phases. The activity *could* be entirely preprogrammed and integrated, or it could be organized in some very different way.

Thus the act of temporal induction, and hence the description in terms of a hierarchy, come from *us*, and not necessarily from the system that we are describing. It is quite possible that animate behavioral systems are organized in ways (e.g., ultra-high parallelism) that are not compatible with our traditional habits of induction and part-whole analysis. Certainly some such conclusion is suggested by the chronic inability of the "soft" sciences to produce powerful formalisms and theories.

A. Branch Points and Information

One of the major problems in induction is what to do with event subsequences that are similar but not identical. An important solution is the use of branch points to allow some elements of a sequence to be equated while others remain distinct. For example, suppose that a system has been observed to emit activities A, B, X, and Y, in the following sequence:

$$\cdots \text{ABXABYABYABXABYABXABXABXABY} \cdots$$

We might well represent this system by a finite-state device as shown in Fig. 1. Here, the state after the emission of B constitutes a branch point, where the system "decides" whether to emit an X or a Y. This use of the word "decides" is critically important. It is a prime example of a behavioral imputation that need not correspond to any mechanism actually used by the system we are describing. In other words, when our inductive analysis leads us to postulate a branch point, we also postulate a decision process. Furthermore, we inevitably go on to ask *on what basis* the decision was made. We ask what *information* goes into determining the choice at the branch point. For example, our finite-state machine in Fig. 1 becomes more understandable if we assert that after emitting a B, it reads a room thermometer; if the temperature is above 72°F the machine emits an X, otherwise it emits a Y. Thus we identify the influence of *information* with (apparent) *choice*. This is, of course, the fundamental intuition of formal information theory; we see here that it is just as fundamental in understanding the organization of behavior.

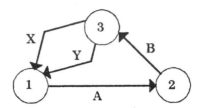

Fig. 1. A finite state generator

Sometimes it is the apparent seeking of information that leads us to postulate a branch point, rather than the other way around. For example, when we see a cat carefully scanning a ledge before jumping onto it, we assume that he is deciding precisely how he can execute the jump, if at all.

Although the postulation of branch points does not force a hierarchical organization (as the finite-state representation demonstrates), the two are very closely related. One simple way of seeing this is to think of a behavioral "parsing tree" as shown in Fig. 2 for the sequence of As, Bs, Xs and Ys given previously: It is extremely convenient to imagine that there is some entity, some *executive decision process*, associated at each branch point, and that this executive entity supervises the activities that are found below it. In the case of an "OR" branching, the executive of course makes the decision as to which branch should be taken. In the case of an "AND" branching, the executive at least decides when one phase should end and the next commence (this can be a nontrivial problem in the control of a complex coordinated activity, such as eating food with a fork).

These executive decision processes, or "executives" for short, which we seem inescapably to associate with points of choice or information intake, are exceedingly important elements in the representation of behavioral systems. The concept of executives will figure importantly in many of the remaining sections of this chapter.

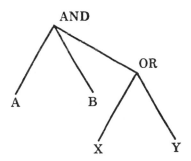

Fig. 2.   A behavioral "parsing" tree.

B. Spheres of Influence

Once we have postulated a hierarchy of executives, it is natural to think of them in terms of the managerial structure of a human organization (see Fig. 3). While there are a number of inadequacies to this metaphor, it can be quite instructive. We think of a human executive as having a certain "sphere of influence". This includes the agents "below" him whose work he controls, and the supervisors "above" him who specify and evaluate his own work. It is important to note that the executive's world-- that is, his sources of information--is *local*, being restricted to the nearby realms above and below him. Of course, there is no precise definition of "local"; what is important is that some information is harder for the executive to come by than other information.

To give an important example, let us consider the case of a man sitting in his living room watching television who suddenly desires a can of beer. At some peripherally conscious level, he realizes that he must get up, go into the

ORGANIZATIONAL CHART OF GENERAL MOTORS

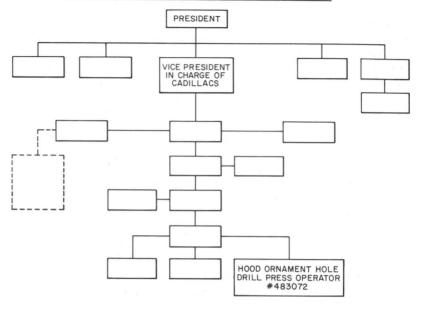

ORGANIZATIONAL CHART OF BEER DRINKER

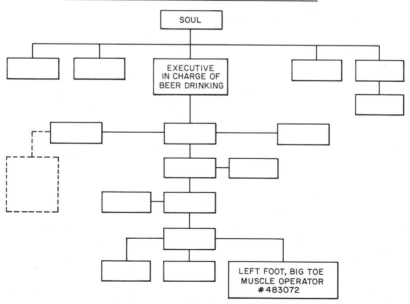

Fig. 3. Organizational charts.

kitchen, and open the refrigerator in order to get a can of beer. In order to get up, he calls on a skilled activity involving placing both feet on the floor, bending forward at the waist, placing his hands on the arms of the chair, etc. In order to place a foot on the floor, specific neural circuits are used, containing internal feedback loops to ensure smooth control of the muscles. Now, what interests us is that the near-conscious executive has not the slightest idea of how the muscles are moved, while the muscular circuits have not the slightest idea of the desirability of beer. (By a valid analogy, a corporation vice-president and a laborer for the corporation have no idea of each other's tasks--see Fig. 3.) Putting this in terms of information and decisions, we can say that the near-conscious planner is not capable of making any decisions on the basis of signals from individual muscles, and the muscular control circuits are not capable of making any choices based on needs or knowledge involving beer.

Many of the hardest problems in describing or programming coordinated behavior arise from precisely such disjoint spheres of influence. For example, consider a man walking down a city street. At one level he decides to look at a particular building, but his eyes were already moving in some other direction for a different reason, and the building in question is out of sight because he has just turned a corner. Such problems of coordination are basic to any behavioral system which is sufficiently ramified to contain executives with nonintersecting spheres of influence. We will return to the matter of coordination after examining one more fundamental notion.

C. Goals

Perhaps the most tenuous concept involved in the description of behavior is that of "goal". Even more than the other notions which we have discussed, the idea of a "goal" is clearly a descriptive artifact. (A clockwork is an example of a complex behavioral system which ticks along perfectly well with no goals driving any of its gears or pinions.) We have found no single answer to the question of the proper role for the concept of goals, but we are

beginning to have some ideas as to where it fits into the
scheme of things.

If we consider our hierarchy of executives, we realize
that the *administrative* tasks performed by these entities
(tasks such as keeping track of which subordinates are
doing what) are distinct from the overall task of the
system.   That is, the manager of a steel mill pushes papers,
but his ultimate responsibility is to produce steel.   We may
suggest that the notion of "goal" arises precisely when we
have such a separation between an ultimate responsibility
and the administrative work required to meet that
responsibility.   In straightforward behaving systems, where
there is no such separation, we do not need to postulate
goals.   For example, the engine of an automobile drives the
wheels, period -- we do not need to say that it has the goal
of driving the wheels.

To take an example at the opposite end of the spectrum,
suppose that a man decides to discover a cure for cancer by
next February.   Here we have the ultimate separation
between the end-product of a system and the procedure for
obtaining it, namely there is no known procedure for
obtaining it.   In this case, the only useful description of
the man's behavior is in terms of a goal.

It is common to talk of goals of states.   Even in the
cancer example, such a notion seems artificial: the man's
goal is to *do something*, namely discover a cure, not to be
in the state of having discovered a cure.   Also, we may
think of organisms whose behavior is commonly described
in terms of tropisms: the worm's goal is to move toward
water, away from light.   Here we may salvage the notion of
state by speaking in terms of gradients, but we should be
aware that we are embalming time or space derivatives in
what is supposedly a static description.   Thus it is unduly
restrictive to think of goals only in terms of states.

Goals, too, are said to be things that are desirable.
What does "desirability" mean?   We suspect that this
concept can be usefully related to that of *expenditure of
resources*.   Suppose that on a Sunday a man has to choose
between going fishing or mowing the lawn.   We observe
him to be packing up his fishing gear.   We then say that
he has selected the goal of doing some fishing, this being
(therefore) the more desirable alternative.   If it had been

possible for the man to do both activities that day (i.e., if he had had greater resoures), there would have been no need for a choice, and the description in terms of goals would have been much less useful. Thus ultimately the notion of goal brings us right back to the notion of branching, of decision.

III. CONFLICTS

A. Resource Conflicts

As the foregoing discussion indicates, there is a close connection between decisions and limitations of resources. If a system had unlimited resources of all sorts, it would still have to make decisions involving coordination (see the next two sections), but many of its organizational problems would disappear.

One interesting example of resource limitation is the phenomenon of focal attention in animal and human cognitive behavior. It would seem that higher animals have enough equipment that they ought to be able to think about several things at once (not to mention simultaneously seeing, hearing, and feeling many different things), yet apparently we are constrained to concentrate on only one or two things at a time. This simple restriction generates a great deal of the complexity of our cognitive behavior. Should we conclude that focal attention is a design flaw in the cognitive systems of higher animals? Not at all, for without it we could never make the inductions (of causal relations, sensory similarity, etc.) that allow us to perceive an orderly world instead of a chaotic mess of sensations. The point is simply that the filtering provided by focal attention has a high cost (in organizational terms) as well as a high value.

The seriality enforced by focal attention seems a wild extravagance of resources when compared with the serial limitations of present-day computers. The strict seriality of most modern computers has tainted many of our notions about the description of behavioral systems. For example, when a process has "AND-ed" subprocesses, we tend to think of them as sequential steps; when a process has "OR-ed" subprocesses, we worry about the order in which they should be tried until one of them succeeds. These primary concerns of the programmer are in fact artifacts of serial processing in our computers. In human managerial systems, and in biological nervous systems, there is ample opportunity for simultaneous activity among processes at the same level. In such cases, the notions of "AND-ed" or "OR-ed" subprocesses merge into each other, and we must find new bases for describing the activity of the supervisor. The next two sections suggest some principles that may be useful.

An important fact about resource conflict is that it may cut across the sphere-of-influence boundaries of individual local executives. For example, no matter what the hierarchical relationship of various processes that may wish to move our eyes, we have only one set of eyes and all visual processes must compete for them. It follows that the entity which allocates such a resource (i.e., one which regulates competition among various processes) must have a sphere of influence that encompasses all of those processes; in other words, it must become a *global* decision maker. This seems to us to be an extraordinarily powerful conclusion. It seems to mean that a system, no matter how homogeneous its elements (e.g., a nervous system), cannot have a homogeneous *behavioral* structure if it contains conflict over resources. There must be some mechanism that allows the attainment and enforcement of a global decision as to the allocation of the resources. We might even suggest that, according to this argument, the appearance of a unitary "mind" is inevitable (albeit at the level of behavioral description) in any system with a high ratio of potential behaviors to bodily resources. D. Bobrow & Norman (Chapter 5) make a similar argument for central control.

B. Condition Conflicts

More general than resource conflicts are the problems that arise when two executives for independent subprocesses have incompatible requirements as to the state of the world. To air condition your house, the windows must be closed; to ventilate it, they must be open; therefore you cannot do both at once.

In an algorithmically behaving system, especially a sequential one, the initial design of the system assures us in advance that the preconditions for a given subprocess will be met at the time the process is called for. The more adaptive a system becomes, the more its organization must explicitly cope with the meeting of preconditions before a subprocess can be unleashed. Perhaps the ultimate of such organization is a collection of independent "demons", which are subprocesses that themselves actively monitor their preconditions, and autonomously commence their activity as soon as their conditions are met. This "pandemonium" organization is powerful because of its inherent parallelism, but in most cases it must be combined with some sort of supervisory mechanism which will provide the requisite administrative (global) control. In order to see how such hybrid organizations function, we must gain an understanding of some of the more basic elements of the condition conflict problem.

We often think of "conditions" in terms of predicates which are either true or false. There are a number of reasons why this conception is inadequate. Many conditions (such as spatial position) take on a range of values, which may well be continuous. In many cases it is worthwhile to consider both the value of some measurable quantity (e.g., intensity of a stimulus) and its time derivative; this complicates the specification of a condition involving such a quantity. Often in real systems, the value of a condition can be obtained only to some degree of certainty less than 100%; in such cases there must be a balance between overhead of ascertaining the condition and the chance of making an erroneous decision. Even worse is the problem of the possible variation in a condition over time. That is, the system cannot afford to monitor all conditions at all times, but conditions may have changed in the interval since they were last observed. This situation has come to be known as the "frame problem"; clearly it implies that conditions must be assigned "expected truth values", rather than being represented as predicates which are either true or false.

These kinds of problems are compounded whenever the

system takes any overt actions, because then it produces
some not-wholly-predictable change in the world. In
general, the possibility that any one subprocess will change
the preconditions for any other (either favorably or
unfavorably) can be computed only in terms of expected
probability, since a system has only a partial knowledge of
the world, and only limited time to spend predicting the
consequences of its actions. Of course, it is precisely this
sort of uncertainty which underlies the importance of
sensory feedback. If you want to know whether or not your
elbow is resting in your coffee cup, don't figure it out --
take a look. Even better, have passive sensors which can
interrupt an action if it results in the placement of your
elbow in the coffee.

The notion of "interrupt" relates back to the idea of a
"demon" silently watching until a certain condition is met,
but it further implies the power of one subprocess to halt
or at least influence another. Once this vital concept is
allowed, our intuitive ability to comprehend the control
organization of a complex behavioral system goes from poor
to abysmal. This is just the point at which we would like
to have a workable mathematics of behavioral organization!

C. Temporal Organization

The problems of condition conflict can be looked at
from a temporal as well as from a logical point of view.
In a sequential system, each subprocess is invoked only
when the previous one is complete, at the behest of the
administering executive. In a pandemonium system, the
temporal interaction is more complex, with the demons
"waiting" in some kind of limbo status until they get an
opportunity to perform, perhaps interrupting some other
demons in the process. In all of this there is still one
element lacking: What sets the pace, what determines the
global temporal organization of events?

In many computer programs, the question of pace is
totally irrelevant. For example, suppose we are given the
mathematical relation $X = (Y+2*Z)^2$. This relation is
inherently atemporal. Now consider a sequential program
for computing X in terms of Y and Z:

(1) X ← Z

(2) X ← 2*X

(3) X ← Y+X

(4) X ← X*X

It does not matter how fast this program is run. All that matters is that the steps be performed in order; this is what determines the equivalence between the program and the formula.

The situation is entirely different for a system that must interact with the real world. If any one subprocess has a temporal extension, then it sets a pace, and all other subprocesses must be placed in some temporal relationship to it. In such a case, the question "What is time?" moves from the philosophical to the practical realm without losing much of its enigmatic character.

We can think of several ways of achieving temporal coordination among processes, each with its advantages and disadvantages. It is possible to define a global time scale, "clock time", against which all activities are mapped out. It is possible to specify events in *relative* time, e.g., B happens five seconds after A, but C is temporally independent. It is possible to control a process in terms of the *rate* at which it proceeds. It is possible to regard time as one of the preconditions to the commencement or branching of a process, e.g., one subprocess could take a certain branch if another subprocess had run for such and such a period, or if the clock time were such and such. No one of these devices is adequate for all purposes, and certainly all are used in effecting the time coordination of human affairs.

We feel that time is less understood relative to its importance than any other aspect of behavioral organization. This is especially true in regard to *simultaneous* processes, which are just beginning to receive formal study. For example, the notions of *monitoring*, and of the *supervision* of one process by another are most clearly exemplified when the supervisor and the supervisee are functioning at the same time. Clearly these and similar concepts are crucial to the organization of process control.

IV. EXECUTIVE FUNCTIONS

A. Executive Bookkeeping

One of the functions of an executive is to keep track of what is going on among its subordinate processes. In current programming systems, subprocesses are usually run sequentially, they terminate of their own accord, and their success or failure is evaluated only after they terminate. Even in so straightforward a case, the executive may require considerable bookkeeping in order to keep track of what has and what has not been done. The problem grows very complicated if the system is to learn anything from attempts which fail. There is the problem of computing which portion of the acquired hard-knock experience was a function of the particular approach that was tried, and which portion is inevitable in any further approach that might be tried. Ultimately, this is a form of the frame problem, solvable only by estimation.

The notions of "success" and "failure" should be treated gingerly, since we should like to distinguish between goals which are explicit to the executive process, versus those which are implicit in the organization of the system (e.g., in our little program for computing $(Y+2*Z)^2$, all goals are implicit, and the executive has only to make sure that the steps are performed in the proper sequence, since they automatically "succeed" and "terminate"). Of course, it is even harder to define when a process is "succeeding" or "failing", in terms of a measure of *progress*, yet this must be done in any system where processes cannot be expected to terminate themselves automatically (e.g., the search for an item in a huge memory store).

The notion of "backtracking" in case of a failure is subject to complexities, even if case failure is well defined, and even disregarding the problem of learning something from the failure. It presupposes that there is in fact a record somewhere of what was being tried, which is not automatically the case in a pandemonium-like system. Also, the problem of diagnosing where to place responsibility for the failure may be essentially insoluble in cases where the chain of command passes through several disjoint spheres of influence. For example, if our beer-seeking television watcher finds that he cannot move his foot (perhaps it is asleep), the analysis of the situation and corrective action must be made at a very much higher level than that at which the failure actually occurred. Thus recovering from a failure may be a challenging exercise in both bookkeeping and decision-making finesse.

B. Executive Decision-Making

Given that the executive can keep track of what its subordinates are doing, in most systems it must allocate "processing power" or some other resource to them on a merit basis. Presumably the most meritorious course of action is that which will produce the best or most results with the lowest expenditure of resource (including time). Of course, the question is how the executive is to know ahead of time, in a non-algorithm-like system, how to estimate the effectiveness and expense of the various alternatives that are presented to it. It is difficult to define how an executive can predict or estimate the behavior of a subprocess without actually carrying out the execution of the subprocess.

We should also mention that the very generation of alternative subprocesses may be a task that consumes non-negligible resources. For example, if an animal is confronted with a visual scene, it must match that scene with long-term memory in order to draw out hypotheses by which it may recognize parts of the scene. This match-search of memory is a major part of the recognition process, and the system obviously cannot afford to draw all possible hypotheses out of memory before testing any of them. Thus part of the executive responsibility is to generate new potential subprocesses in a manner which is efficient, as well as efficiently managing the subprocesses which have already been proposed.

This sort of executive decision-making is perhaps the crux of efficient behavior. At the same time, it is relatively simple conceptually (if stated as a choice among alternatives), and relatively well studied by traditional means (e.g., statistical decision theory). Therefore we will go no further into the mechanisms of decision-making here, since our object is to consider the structure of the behavioral system as a whole. We should note that while it might be relatively simple to enumerate the criteria for any individual decision, it is usually not so simple to specify how such decisions should interact, how executives should coordinate and decide priority among themselves, what spheres of influence should be open to each executive, and so forth.

C. Statistical Information

The description of a behavioral system must take into account the system's behavior over a large class of similar situations; that is, it is essentially an inductive process. For example, suppose that you are introduced to a person, and he reacts moodily to your attempts to talk about football. From this one event, you have no idea whether the problem is that he hates football, or that he hates introductions, or that he took an instant dislike to you, or that he was having a bad day, or that he is generally a surly person. These possibilities can be distinguished only by observing him in a number of similar situations. It is easy to see that the same sort of procedure is necessary for arriving at the proper description of any complex behavioral system.

We would like to emphasize that the information gathered in such experimentation with a behavioral system is *statistical* in nature, and that therefore the selection of a model of a behavioral system is closely connected to the statistics of its responses to typical inputs. This fact has been implicit in everything we have said about alternative organizations of systems. For example, if subprocess B is *always* both desirable and possible after the execution of subprocess A, then the best organization is to make them sequential steps under some larger process. If the applicability of B depends on some particular set of conditions, it might be best to provide a test of those conditions, with the execution of B being dependent on this test. If B is only rarely applicable, or if the circumstances of its applicability are not readily predictable from tests, it might be best to establish B as a "demon" which independently waits and watches for its opportunity to proceed.

Thus the proper organization of a particular behavior is entirely dependent on the statistical peculiarities of the task at hand. This is true both of the global organization, and of the details of control throughout the system. Furthermore, there are some problems, such as the handling of the "frame problem", which have solutions *only* in terms of a statistical conformity of the system to its informational environment. Perhaps this ubiquitous influence of the statistical properties of the task is the most important general principle that can be stated about the organization of behavioral systems.

V. THE RELATIVITY OF BEHAVIORAL DESCRIPTION

In the foregoing sections we have discussed some concepts that seem central to the description of behavior in a large variety of systems, both animate and mechanical. We can now pose a question that was hinted at in our introduction: Can these concepts be developed into a mathematics of behavioral systems, in the sense that traditional mathematics provides concepts for the description of physical systems? We suspect that the answer to this question is no.

Why not? Let us consider first the application of calculus to the description of a physical system, such as an electronic circuit. Perhaps we have a capacitor, and we wish to consider its rate of discharge. Now, no person with an expensive technical education will find any ambiguity in the phrase "rate of discharge", nor will he have any hesitation in applying this notion to the description of the capacitor, nor will he have the slightest difficulty formalizing it in terms of differential calculus.

Now let us consider the problem of describing a behavioral system, such as our household cat in the process of tracking down a mouse. One might expect that the notion of the cat's "behavior" at a given moment should be as unambiguous as the notion of the capacitor's "rate of discharge", but it is easy to show that this is not so. With only a little reflection we can think of at least five different statements of the cat's behavior:

(1)   He is putting one paw before another in such-
       and-such a manner.
(2)   He is slinking.
(3)   He is stalking the mouse that he sees.
(4)   He is hunting.
(5)   He is trying to secure a meal, to satisfy his
       hunger.

Whatever the merits or demerits of these particular five
descriptions, it should be clear that in the case of
behavioral systems there is simply nothing like the
straightforward correspondence between phenomenon and
description that exists in the case of physical systems and
their mathematical representations.

A "behavior" is not a well-defined thing like a
numerical quantity; rather it is a selective account of an
event that is defined only by certain decisions on the part
of us, its describers. Some of these decisions are: Are we
describing a particular behavioral episode, or must we
account for all potential episodes? Are we interested in the
utmost details of the behavior, which are different each
time, or are we merely indicating the general outline of
events, which is what typifies the behavior for us? Must
we determine the factors which might motivate or evoke
the behavior? Do we include the potential interaction of
the behavior with an observer (i.e., "if you do so-and-so,
the system will respond so-and-so")?

Questions like these, answered implicitly in one way or
another, determine what "behavior" we are considering, and
therefore select the form of our behavioral description from
a wide range of possibilities. The multiplicity of possible
ways of delineating a behavior accounts for the
impossibility of any one-to-one mapping between a
behavioral event and its description. Perhaps another
example will clarify this point a bit.

Consider a desk calculator of the sort that is 100%
mechanical. A set of blueprints for such a machine will
tell us all that we can possibly know about its structure; in
fact, they should allow us to build a working copy of the
device ourselves. Paradoxically enough, the blueprints -- a
complete structural description -- fail to answer the
questions that we must be able to ask of a *behavioral*

description: "What does this thing *do*?", "*How* does it *work*?". The blueprints, for example, tell us nothing at all about how the calculator performs division, and when we come to create a behavioral description of the process of division, we find that this is not yet well defined. Are we referring to the sequence of manipulations by which the operator causes the machine to divide two numbers? Are we describing the mathematical operations of repeated subtraction, rounding, and so on that the calculator performs in order to accomplish division? Are we detailing the sequence of ratchet, pinion, and shaft movements that are the actual events occurring within the machine? Even in this simple example of mechanical behavior, where we are given a complete structural description of the system, there *still* is no unique behavioral description of the system's primary functional operations!

In summary, there is no straightforward, absolute, canonical, or true description of a behavioral system; all behavioral descriptions are relative to a particular set of questions they are intended to answer. Of course, it might be said trivially that *any* description has validity only in terms of the questions it is intended to answer, but there is a real issue here beyond mere toying with words. In the case of the traditional mathematical description of physical systems, the categories of possible questions are generally agreed on, broadly applicable, and therefore implicit in the procedures we have learned for viewing the physical world. The very existence of traditional mathematics speaks for its general, implicit rules of applicability. When it comes to the description of behavioral systems, as we have seen, there is no general rule that will govern the application of descriptive concepts even in the simplest of cases (which is merely to say that even the simplest cases of behavior are not at all simple). Thus there can be no "mathematics" of behavioral systems in the traditional sense of a conceptual system which can be applied descriptively in an automatic, canonical way.

VI. SUMMARY

We have explored the origins and interconnections of some of the concepts that are generally useful in describing the behavior of animals and machines: hierarchical organization of processes, branch points, information, spheres of influence, goals, resource conflicts, condition conflicts, temporal organization, executive bookkeeping, executive decision-making, and statistical information. We have seen that by and large these concepts must be viewed as descriptive artifices rather than as mechanisms by which the observed behavior is brought about (unless, of course, the behavior itself is artificial, as in the case of computer programs in which these concepts are intentionally used as mechanisms). Finally, we have considered the problem of turning this set of concepts into a full-scale mathematics of behavioral systems, finding that this may be impossible owing to the lack of any generally applicable criteria for delimiting what "a behavior" is.

Our purpose here has not been to prove rigorously one point or another, but rather to encourage the behavioral scientist to pause and reflect a bit on the conceptual tools of his trade. The tools are good ones and, if properly used, can lead us to a genuine understanding of complex behavioral systems that have heretofore defied description.

ACKNOWLEDGEMENT

A friend of Jaime Carbonell's whose name unfortunately does not appear in the table of contents to this volume is E. William Merriam. In preparing for the Carbonell conference, Bill and I co-authored not one but three papers, describing the theoretical bases of the simulated robot cognitive system that he and I have been working on for several years. One of these papers was presented by Bill at the conference, but ultimately we did not judge any of them to be definitive enough to be published. I and Jaime's friends are extremely grateful to Bill for the effort he expended on behalf of this conference, and of course I have greatly benefited from his help and expertise as a co-worker over a number of years.

SYSTEMATIC UNDERSTANDING:

Synthesis, Analysis, and Contingent Knowledge
in
Specialized Understanding Systems

Robert J. Bobrow
and
John Seely Brown
Bolt Beranek and Newman
Cambridge, Massachusetts

I. INTRODUCTION

In the design of an AI program that understands a domain of knowledge and efficiently copes with problems in that domain, careful attention must be paid to the choice of representations for different bodies of knowledge. Clearly, the best representation for a body of knowledge depends on how that knowledge is to be used by the program, and thus better characterization of the uses of knowledge is likely to lead to better ways of designing knowledge representations. In this chapter we present *the SCA model*, a framework for describing the structure of "conceptually efficient"[1] understanding programs, based on a characterization of three fundamentally different ways in which knowledge is used in such programs.

Even though the SCA model is not fully developed, we feel that it can be of use both to those designing understanding programs, and to those who wish to study existing programs to develop insights into different approaches to representing knowledge.

The version of the SCA model described in this chapter applies to a class of programs that we refer to as *specialists* or *expert understanders*. These programs understand the world in the sense that they can take in a collection of data describing some situation, and then answer questions about that situation. The programs are referred to as specialists because they are only able to deal with a limited class of situations, and can only answer questions in some limited domain of specialization.

We are still in the process of developing the SCA model, and many ideas are still in the form of speculations and intuitions. In order to present the essential aspects of the SCA model in the clearest possible form, we have described a simplified version of the model which does not deal with a number of crucial issues in the design of understanding systems. In particular, though the systems to be discussed "learn" about their environment in the simple sense of taking in descriptions of the current state of the environment, they do not learn how to improve their performance, nor do they extend their initial knowledge about those properties of their environment which hold in

---

[1]D. Bobrow has suggested this term as a way of distinguishing programs whose speed is due to their way of conceptualizing the structure of the problem domain, as opposed to ones which operate rapidly because of clever but nongeneralizable coding tricks.

all states. Additionally, though we believe strongly that complex problems in understanding will be solved by programs built from many specialized modules that communicate and interact within a complex control structure, the model we describe consists of two programs that interact by simply creating and accessing a common data structure.

Part of our intent in articulating the SCA model is to present a point of view on the design of expert understanders which provides insight into how knowledge of the expert's domain can be effectively used to design a conceptually efficient understander. The SCA model provides a framework for studying the structural similarities of a variety of superficially dissimilar high-performance understanding systems including perceptual understanding systems, natural language data base management systems and question-answering systems. It also provides a basis for discussing many current AI research issues such as the meaning of analogical representations, the pros and cons of different inference schemes (e.g., computational, rule-driven,...), ad hoc versus general knowledge representations, and the integration of "non-AI tools" (such as simulation) into AI systems.

II. THE SCA MODEL

A. Brief Description of the Model

For the purpose of providing an overview of this model, let us consider a hypothetical expert understander program which obtains information about the world in the form of a collection of basic uninterpreted "sensations" or raw input data. The expert program must answer questions about the world on the basis of this raw data. Much like a human expert, our hypothetical expert uses its specialized knowledge to combine, organize and augment this raw data, and thereby synthesize a world model or *contingent knowledge structure* (CKS) whose organization and content are substantially richer than the collection of raw input data

In its simplest form this CKS might just be a data base

of assertions describing the expert's current knowledge. In general, however, the CKS is a *complex data structure* that represents the expert's current model of the world. This model may include analog representations of some aspects of the world, as well as semantic networks and other knowledge representations.

The SCA model extends the concept of actively building a world model from raw input data, and specifies that an understander should be designed as two separate expert programs, one to *synthesize* a CKS from the collection of raw input data and the other to *analyze* the information in the CKS in order to answer the questions posed to the remainder of the understander. The raw input data are the result of the operation of other programs or input devices, and are not usually represented in terms of concepts that are directly usable by the understander. The first expert, or *synthesist*, converts this raw input data into a form suitable for the remainder of the understander to act on.

In addition to simply augmenting the information contained in the raw data, and reorganizing this data, the synthesis operation often changes the elementary concepts in which the information is expressed (e.g., changing from a representation of a visual scene as an array of intensities to a collection of boundary lines, or from a collection of boundary lines to a set of three-dimensional objects).

The second expert, or *analyst*, retrieves knowledge explicitly represented in the CKS, and uses world knowledge to infer facts not explicitly contained in the CKS, perhaps using procedures as complex as general theorem provers.

The effectiveness of the total understander depends critically on its designer's ability (i) to provide the correct balance of capabilities (and computational load) between the synthesist and the analyst, (ii) to design a class of CKSs which represent just those aspects of the world state needed to facilitate the operation of the analyst, and which can be directly represented as efficient combinations of procedures and data structures, and (iii) to use special purpose techniques to enable the synthesist to fill in the CKS in an efficient way.

Before describing some of the theoretical issues that follow from this seemingly straightforward decomposition of information processing tasks, we provide several instances of systems that exemplify the model.

B. A Perceptual Understanding Example

Consider the problem of building an understander for a blocks world which might combine some features in programs such as described by Winston (1975) and Winograd (1972). This understander might answer questions such as:

(a) Where is the tallest block?
(b) What blocks are to the left of the block at (10,25,17)?
(c) What block is the closest neighbor on the left of the block at (11,5,3)?
(d) How much of the top face of the tallest cube is covered?

We assume as input a set of *line segments on a plane* which correspond to a camera image of the boundaries of three-dimensional blocks in a scene such as that shown in Fig. 1.

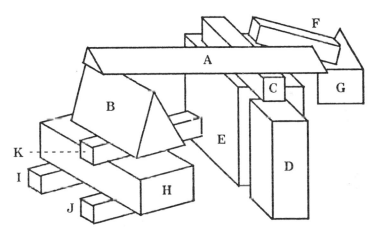

Fig. 1. A blocks world scene to be analyzed.

Notice that the questions deal with relations among three-dimensional items such as blocks, faces, distances, shapes, etc., while the input data only refers to two-dimensional lines and their endpoints. To cope with this difficulty, an understander built according to the SCA model would have a synthesist which takes a set of lines and produces a CKS representing a configuration of blocks that would account for the lines. The analyst portion of the understander would answer questions about the configuration of blocks by analyzing the structure and contents of the CKS which represents the configuration.

The synthesist determines a physically possible configuration of blocks that accounts for the perception of the line segments given as input. It then creates a data structure that both represents the blocks and simplifies the operations performed by the analyst. For simplicity of description we have finessed the problem of *ambiguity*, the possibility that several configurations of blocks might account for the given boundary lines. In general, the resolution of ambiguity requires additional interaction and feedback between the analyst and the synthesist, and possible feedback paths from the programs that invoked the understander.

The synthesist would contain, implicitly or explicitly, much information about the world of the understander, such as:

(a) the types of objects in the world--planar faced polyhedral blocks, no wires or painted lines, etc.
(b) the laws governing the possible arrangements of blocks--blocks must be supported, blocks cannot interpenetrate, there is a planar supporting surface (the table), etc.
(c) the perceptual system--perspective projection of the boundaries of objects on a plane, objects occlude one another, etc.

A direct way of using this knowledge is for the synthesist to propose the presence of a particular type of block, and use this hypothesis to account for lines in the input [as was done by Roberts (1963)]. Unfortunately, this technique fails catastrophically when the number of possible types of blocks is very large or when the scene gets complex and has many blocks occluding one another.

This technique is an example of an *expectation-* or *hypothesis-driven* (or *top-down*) synthesist which has a strong model for the structure it expects to synthesize, and attempts to use that model to account for the input data. The choice of terminology is deliberately close to that used by D. Bobrow & Norman (Chapter 5), since we believe that their notion of "accounting for all of the input data" is closely related to our conception of synthesis. The use of the term "top-down" is intended to suggest the overlap of the concepts of parsing and synthesis (as will be more fully developed in the example that deals with the LUNAR system).

A better way of using the available world knowledge is the constraint satisfaction technique described by Waltz (1975). Its goal is to find one or more *global interpretations* of the set of line segments as boundary lines of blocks consistent with *local constraints* dependent on properties of junctions of line segments. Each junction is assigned a set of interpretations in terms of the possible properties of the blocks which meet at the junction. Waltz's algorithm then finds sets of global interpretations of the line segments which are consistent with the local interpretations. Given the resulting grouping of line segments into clumps corresponding to single blocks, and information about the position of the camera relative to the blocks, a synthesist could determine the position and shape of each block (making reasonable assumptions about invisible faces). This is an example of a *data-driven* or *bottom-up* synthesist, where the characteristics of the input data suggest the structure and content of the CKS used to account for the data.

To complete our sketch of how an understander might be organized according to the SCA model, we will give a brief indication of possible structure and function of the modules of the understander corresponding to the CKS and analyst.

A single CKS in this blocks world might consist of linked list structures describing relations among the planes, boundary lines and points composing each block. These block descriptions could be indexed by a three-dimensional array, with the intended meaning being that if block B is pointed to by A(I,J,K), then block B occupies (part of) the volume of an imaginary cube centered at the coordinates (I,J,K). This indexing would simplify the answering of the questions "what is the first block to the left (right) of block B?". Relations like *support* and *to the left of* would not be explicitly included in the structure, so that if a block is moved it would be easy to update the structure.

The analyst for this blocks world might combine procedural specialists for answering simple questions (e.g., "what is the set of blocks to the left of block B?", "what is the height of a block B?", etc.) with a mechanism for converting complex English questions to compositions of simple questions, and an interpreter for evaluating such compositions.

C. A view of SOPHIE's Procedures for Answering Hypothetical Questions

That part of the SOPHIE system which answers hypothetical questions (Chapter 11 by Brown & Burton) fits nicely into the SCA model. Its input consists of hypothetical statements about the setting of various controls and the condition or values of various components of a fixed electronic circuit. Using this information it must be able to answer any question about the implied state of the circuit that someone with a volt-ohm-meter and some electronic knowledge could answer--for example, the voltage between any two nodes, or the current through any component.

The hypothetical-question answerer is an example of a useful class of SCA understanders we call the "systems understanders". A *system understander* is a program which uses models for individual world states as a basis for understanding a *system* or constrained collection of world states (e.g., the set of states of the circuit in SOPHIE). The defining property of a system understander is that it only handles questions about the properties and relations among the entities in a single state, not questions concerning relations between the properties of entities in different states. Our experience and theoretical work indicates that one can build extremely efficient understanders if the range of acceptable questions is restricted in such a manner. Although no problem solving or comparison of different states can be done *within* a system understander, it can be used as an efficient component of a more complete system, by the techniques illustrated in SOPHIE (Chapter 11).

Three separate but interdependent modules must be designed for any system understander:

(a) the *state representation scheme* (class of CKSs): a class of data structures and associated procedures designed to represent arbitrary *individual* states in the world,

(b) the *query evaluators* (analyst): procedures which use the information in a single state representation to answer questions about the properties of the entities and their relations in the state,

(c) the *state representation generator* (synthesist): a procedure which generates a state representation given the input to the understander.

SOPHIE's circuit simulator, along with the procedures for modifying the circuit description it uses, constitutes a state generator. It takes statements about the setting of circuit controls and the characteristics of the circuit components and generates a more useful description of the state of the circuit, in the form of a table of voltages at the nodes of the circuit. This table, combined with SOPHIE's semantic network, which describes the connectivity of the circuit components and the values of each component, forms the state representation.

Factoring the state representation into the voltage table and semantic network allows the state generator to produce a new representation by simply generating a new voltage table, without having to reproduce unchanging state information (such as connectivity). The representation of the state in terms of voltages has two other useful properties:

(i) it is nonredundant, with all other properties of the state being computed from the voltages and semantic network, so that there is no problem of maintaining consistency when the state representation is updated (see Chapter 1 by D. Bobrow);

(ii) its uniformity simplifies the construction and speeds the operation of the analyst, since the analyst does not need to check what type of information is available, nor how to use it to answer questions.

The circuit simulator is an example of a synthesist designed to create contingent knowledge structures which always have the same form, consisting of a voltage table and a semantic network. Since the laws of electronics provide simultaneous constraints on the voltages to be filled in, and since the circuit contains nonlinear components (transistors, diodes, etc.) and feedback loops, the complete determination of the CKS (including the voltages in the circuit) is a nontrivial task. Unlike the blocks world synthesist, this one does not have to determine the structure of the CKS, only the values of parameters.

Such a "fixed form" synthesist is a good example of an "expectation driven" synthesist--it expects to be able to account for its input data in terms of a CKS of a very limited form, and cannot operate on inputs which cannot be accounted for by such a CKS.

One can view the rules defining the system either as logical axioms giving partial information about the world, or as constraints on the states of the system. By viewing the state generator as a *device for synthesizing a model of a state* meeting certain constraints, rather than as a formal deductive inference mechanism, the designer can construct a highly efficient program by taking advantage of special characteristics of the state generation process:

(a) The structure and components of the output are known when the system is designed, so the control strategy for using information from the system definition can be "compiled in", rather than requiring extensive tests or search based on the information available when the simulator is run.

(b) The simulation need not provide explicit formal proofs of its results. Thus the steps of the computation need not be recorded nor shown to conform to some specified set of rules of inference. During the design phase one can verify that the results of the simulation will always meet the system constraints, and thus omit explicit verification of this fact in the operation of the simulator. This allows one to find computationally efficient characterizations of the system, and use those, rather than the originally given constraints, to generate the states.

(c) A correctly designed simulation can generate all of

the information needed in a state representation "simultaneously" in substantially less time than it would take to generate each of the individual bits and pieces separately. This speed-up is particularly evident in systems with feedback or constraints which establish relations that must hold simultaneously among several properties of a state. In these cases it is impossible to determine one property without determining several others, at least implicitly.

That part of SOPHIE's hypothetical-question answerer seen as the query evaluator consists of several elements: a set of primitive operations for determining components and nodes in the circuit, given their descriptions; operations for measuring electrical properties in the circuit; and a mechanism (essentially the LISP interpreter) for composing these operations to answer a particular question.

The designer should view the query evaluators and state representation as forming an *analog model* of the state by which we mean a device for predicting the results of measurements on the state described by questions in the query language. This emphasizes the need for coordinated design of the data base and query evaluators, taking into account both the elementary question-answering operations and their composition rules. The analog must *act like the state* only in terms of providing the same results to questions in the query language.

D. The LUNAR System--Parsing as an Example of an SCA model

Woods's LUNAR system (Woods, 1973) is an experimental natural language data base management system designed to permit a lunar geologist to access, compare, and evaluate conveniently data on the lunar rock samples from the Apollo 11 mission. The system takes such questions as:

WHAT IS THE AVERAGE CONCENTRATION OF ALUMINUM IN HIGH ALKALI ROCKS?

and provides answers based on a table of over 13,000

analyses of the Apollo 11 samples. LUNAR provides both a clear example of the SCA model and an example of how a complex synthesis procedure can itself be decomposed according to the SCA model.

The synthesis phase of the overall LUNAR system can be seen in the natural language parser and semantic interpreter that convert English sentences to statements in a formal query language. The query language is a generalization of the predicate calculus whose statements can be directly executed on the data base to provide the answers to the corresponding English question. Each individual statement constitutes a CKS.

The analysis phase is implemented by the LUNAR interpreter which consists of the LISP interpreter, augmented with a set of specialized retrieval procedures and predicates. This interpreter extracts information from the fixed data base by executing the CKS as a program.

The two previous examples have shown a type of synthesis that might be called "data synthesis", since it builds a CKS to account for and represent the raw input data. This example shows an example of "goal synthesis", since the synthesizer produces a CKS that represents the goal of the understander in the given context, i.e., the question to be answered. Goal synthesis is found in many understanders, and in particular is found in all understanders that use "natural language" input. In such understanders there is always some portion of the program devoted to converting the natural language input to some other representation which is easier for the program to handle. Not all such systems use the further decomposition described in the next paragraph, and in particular, the natural language section of SOPHIE converts its input directly to an internal semantic representation without help of an explicit intervening syntactic CKS. Such an approach loses some of the power and flexibility that the syntactic CKS-semantic interpretation approach offers, but can be faster when the input is sufficiently limited to be parsed directly into the semantic form.

The top-level synthesis operation in LUNAR is quite complex, and is itself broken down into a syntactic synthesis operation that produces a parse tree as a CKS, and a semantic synthesis operation that obtains its input data by actively applying analysis procedures to extract information from the parse tree. The parsing synthesis uses an augmented transition network parser (Woods, 1970) and heuristic information (including semantics) to produce the "most likely" parse tree for the sentence. This parse tree is a CKS that makes explicit properties such as the logical subject of the sentence; whether the sentence is declarative, imperative or interrogative; whether it contains a negative element, and if so what the scope of the negation is.

The parsing synthesis must use the global structure of the whole sentence to determine its logical subject. The synthesist places the subject in a fixed structural position in the CKS which enables the analysis operations that produce the input for the semantic interpreter to obtain information that depends on the *global* structure while performing only *local* operations on the parse tree. Pulling together scattered pieces of information and providing a simple means for later processes to access them is a major purpose of all synthesis mechanisms, and is an important aspect of why parsing is actually done in question-answering systems.

III. THE ELEMENTS OF THE SCA MODEL

A. The Synthesist

The synthesist describes or accounts for a collection of input data in terms of some *acceptable* structure, which is an instance of a class of structures specified by the designer. It must impose the "best" acceptable structure on the input data (which may already be organized in some fashion). The imposed structure accounts for the original input data in terms of concepts useful to the analyst, and provides the only mechanism through which the analyst is permitted to obtain information about the state of the world. In general, the analyst never has direct access to the information contained in the raw input data. In more elaborate versions of the SCA model the analyst can request the synthesist to synthesize a new CKS based on the current needs of the analyst. This may change the effective information that the analyst obtains from the raw input data, by means of the CKS, but it does not remove the CKS as a necessary intermediary.

There are three interrelated operations that may be performed by the synthesist:

(i) *structure determination*--determine which of the potential structures it knows about (e.g., configurations of blocks, parse trees, etc.) should be used to represent the given raw data,

(ii) *matching*--determine the correspondence between the raw data items and the parameters of the chosen structure, and,

(iii) *augmentation*--determine parameters of the CKS not directly corresponding to raw data items, but which must be chosen to satisfy some set of constraints.

These operations may require a coordinated search through both the entire set of facts and the possibly infinite set of potential structures, in a matching or parsing phase. The synthesist may put the given facts in a canonical form, for example by algebraic simplification, by reduction to a structure composed of semantic primitives (Schank, Chapter 9), or by using hash-coding and search to reduce an expression to a unique internal structure (Reboh & Sacerdoti, 1973). Finally (as in the simulator in SOPHIE), the synthesist may fill in parameters of the structure in a manner determined by the raw input data and a set of constraints. The synthesist may only have to perform one or two of these operations, as when the matching operation and structure determination are trivial but the augmentation operation is difficult (e.g., the synthesist in SOPHIE).

It is often possible to distinguish repositories of knowledge corresponding to an active synthesis agent, a description of the *class* of possible contingent knowledge structures, and a goodness measure which evaluates how well a structure matches up with the raw input data. In the LUNAR parser, for example, the description of the class of possible contingent knowledge structures (as well as rules for searching for the "best" structure) is given as an augmented transition network and the active agent is the general ATN parser. In many cases, however, it is impossible to separate the synthesizing agent from the description of the structures to be synthesized. In these cases the synthesist is implemented as a specialized procedure for instantiating a particular class of structure-- this is often more efficient, though less flexible.

It may seem possible to build in accessing functions in the analyst and do away with synthesis, particularly where the information used by the analyst only depends on a small amount of raw input data. The choice of structure, however, even for a small portion of the raw input data, often depends on the entire collection of raw input data. In these cases one cannot do without the synthesist by simply putting extra processing in the analyst--the extra processing needed is actually the complete synthesis operation. For example, in the blocks world the analyst need only look at a structure determined by a small part of the input to answer questions about a single block, but the very concept of a block depends on a synthesis process that takes into account the entire set of line segments in the scene.

B. Contingent Knowledge Structures

The *contingent knowledge structure* (CKS) is a data structure which represents the understander's knowledge of the state of the world. The CKS forms the interface between the synthesist and analyst, and thus determines the way in which the characteristics of the world are interpreted by the specialized question-answering and problem-solving capabilities of the analyst. Because of this the capability and efficiency of an SCA understander depends critically on the design of the CKS and its computer representation.

There are two distinct aspects to the design of the CKS:

(a) the conceptual structure which the CKS imposes on the world, and

(b) the data structures and procedures chosen to represent the entities, properties and relations which make up the conceptual structure of a CKS.

Like a language, the class of possible CKSs defines a set of conceptual entities, relations, and structures which determine the way the analyst can most easily express its questions and operate on its model of the world. It determines the distinctions which the analyst can possibly make between states of the world. For example, in the blocks world understander the conceptual entities are blocks with sizes, shapes, positions in three-dimensional space, and relations depending on their positions--rather than corners, connected sequences of line segments, or any other possible structural entities.

Given several alternative conceptual structures for the class of CKSs, it is tempting to choose the "most expressive" one, to facilitate the description of the states of the world and the questions of the analyst. The analyst, however, must be capable of dealing with any CKS within the chosen conceptual framework, and more subtle and complex frameworks require more complicated (therefore usually less efficient) analysts. Thus there is a tradeoff between expressiveness and efficiency, and a good conceptual structure is one that can readily express those properties of the world that are relevant to the analyst, but does not lend itself to describing irrelevant details or impossible states. An example of an unproductively rich conceptual structure for the blocks world CKS would be one that could represent all connected sets of line segments, including collections of lines not corresponding to the boundaries of a set of blocks. Such collections are impossible in the blocks world and hence are irrelevant to the blocks world analyst.

We believe this tradeoff is closely related to one hypothesized by Thompson of Caltech (Thompson, 1972; Randall, 1970; Greenfeld, 1972). Thompson's conjectured "fundamental theorem of information theory" says roughly that given a collection of observations of the world, and a sequence of progressively richer languages, there is an intermediate language in which the descriptions of the observations carry the greatest information. This most informative language is not rich enough to express all the details of every observation--the concepts that make up its semantics are broader and more abstract than the details of the observations, and thus it captures the important properties of the observations without allowing the expression of unnecessary irrelevant detail.

By fiat the CKS provides the only mechanism for the analyst to refer to the properties of the world state. The CKS reduces the complexity of the analyst by providing a form of canonical representation for world states--states of the world which can be treated similarly by the analyst are mapped into similar CKS data structures. Thus the conceptual structure must be designed to facilitate both the expression of the questions which the analyst must answer and the operations which the analyst must perform in order to answer the questions.

We believe that in most cases the CKS is best viewed as a *model* for the state of the world, rather than a description of the state, so that the operations of the analyst correspond more to observations of the world than to manipulation of the representation of assertions to determine their implications according to some set of rules of inference. We use the term "observation" in the sense of an operation on the world that produces information as a result, and which does not change the state of the world (this is in accord with the simple version of the SCA model, but is not a general restriction). We do not mean to imply that such observations are simple operations, or that the analyst is simply an information retrieval mechanism that observes what is explicitly present in the CKS.

Much of the understander's knowledge of the state of the world is not contained explicitly in the CKS, but is embedded in a set of tacit agreements between the synthesist and analyst as to the way in which the data structures that form the CKS are to be interpreted. For example:

(a) A linear sequence of links between several items can be interpreted as a transitive relation if the analyst determines whether two items are related by seeing if there is a chain of links joining them.

(b) A set of pointers to objects from elements of a three-dimensional array is sufficient to represent many of the three-dimensional relations among a group of physical objects if both the synthesist and analyst interpret the existence of a pointer from an array element $A(I,J,K)$ to an object O as meaning that O intersects a box centered at coordinates (I,J,K).

(c) A list structure can be used to represent a procedure for answering a question, given a set of rules such as those used to interpret LISP forms.

Given a particular choice of conceptual structure and a set of operations (defined in terms of the conceptual structure) which the analyst must perform, the designer must choose a collection of computer data structures to represent the CKS, which maintains the comprehensibility of the final program and its overall efficiency. The distinctions we draw between the conceptual structure of the CKS, the implementation of the conceptual constructs in terms of programming language constructs, and the eventual implementation of these constructs in terms of machine-level primitive operations are an attempt to deal with the problem that Hayes (1974) poses:

> A representation which appears to be a direct model at one level...may...be itself represented in a descriptive fashion, so that it becomes impossible to describe the overall representation as purely either one or the other. It seems essential, therefore, to use a notion of *level* of representation in attempting to make this distinction precise.

C. The Analyst

An analyst derives information from a CKS in order to answer questions posed by some other process. Informally, if the CKS represents a state of the world viewed from a perspective defined by the conceptual structure of the CKS, then the analyst infers answers to specific questions about the state of the world using information in the CKS and a set of rules. These rules include laws of the world and laws of logical inference, so that the answers provided by the analyst correspond to true propositions about the state of the world. (In more complex versions of the SCA model, the analyst may produce "plausible" results as well as "necessarily true" ones.) The CKS is of necessity a finite collection of information, but the set of questions one can ask about the context usually come from an infinite set defined by linguistic rules. Thus *the analyst is an inference mechanism for bridging the gap between the finite set of the properties of the world which are explicitly represented in the CKS, and the infinite set of valid assertions* (answers to questions) *about the world which are implicitly determined by the CKS.*

The simplest analytic operations consist of the application of compositions of functions and relations to elements of a CKS, for example forming the sum of the lengths of several lines in a geometric structure, or comparing such a sum with the distance between two points. More complicated analytic operations might consist of using the results of such simple operations, along with general world knowledge to deduce further properties of the world state represented by the CKS. *In many cases the understander need not explain or explicitly justify its answers, so explicit logical deduction can be replaced by other forms of inference or computation.* For example, in a CKS which consists in part of a semantic network, properties of an element can be inferred by tracing a path to some other element and then applying simple computational rules to a description of the relations in the path and the properties of objects on the path.

The fact that "John's uncle is Jane's grandfather" could be derived from a chain "John **son-of** Peter **husband-of** Mary **sister-of** Isaac **father-of** Ellen **wife-of** Jack **father-of Jane**", by the application of a set of simple rewriting rules to the sequence "**son-of husband-of sister-of father-of wife-of father-of**".

IV. COMMENTS ON REPRESENTATION ISSUES

A. Universal and Ad Hoc Knowledge Representations

A theory such as the one which we hope underlies our SCA model would make it possible to discuss rationally the representation of knowledge at the interface of the synthesist and analyst, as well as the design and operation of synthesists and analysts for particular problem domains. It would permit us to phrase meaningful questions about the relative merits of different contingent knowledge representations from the point of view of efficiency of synthesis, analysis and original design, and could thus clarify the debate over the relative merits of "general purpose" and "ad hoc" representations of knowledge.

Let us take a superficial look at this debate in terms of the SCA model. Given that all understanding systems must convert their raw input data into data structures used to meet goals of the understander, the more goals that can be met effectively with a single structure, the fewer times must a synthesist be invoked to create another one. Building a single structure to serve many independent goal-oriented procedures may, however, be more difficult than building several different specialized structures. In addition, the improvement in efficiency of the goal-oriented analytic operations brought about by the availability of specialized contingent structures may make up for the extra time spent in building them.

One must also take into account the resources necessary to maintain consistency and compatibility among multiple representations, or allow for the problems of potentially inconsistent actions by different analysts.[2] These problems are particularly difficult to resolve in understanding systems which generate and deal with multiple contexts, such as planning and problem-solving systems.

One must also evaluate the impact of "generalized" and "ad hoc" representations on the problem of designing systems. Clearly, having a single, highly efficient, and effective knowledge representation would substantially reduce the time necessary to design new understanding systems. Even a unified conceptual framework like "the omega order predicate calculus" (as in QLISP) can ease the designer's task. On the other hand, the usefulness of engineering handbooks attests to the fact that an organized collection of specialized structures with capabilities and limitations clearly spelled out can be quite as good a design aid as a single generalized technique.

---

[2]There is certainly evidence that human understanders have inconsistent representations of knowledge, and that they can come to inconsistent conclusions by using different techniques for solving problems or answering questions.

B. Contingent Knowledge Structures and the Antecedent/Consequent Boundary

In building understanding systems with procedural representations of knowledge, there is a serious design problem in the distribution of expertise between the antecedent ("if-added") and consequent ("if-needed") procedures [as in PLANNER (Hewitt, 1972), CONNIVER (Sussman, 1972) or QLISP (Reboh & Sacerdoti, 1973)]. Roughly speaking, if-needed procedures are triggered by the introduction of specific goals and subgoals to be met by the understander, while if-added procedures are triggered by addition of new facts to the contextual knowledge base.

The if-added procedures clearly correspond to the operations of synthesis. It may not be so clear that if-needed procedures correspond to a combination of synthesis and analysis, and not simply to analysis. Simply speaking, the if-added procedures correspond to data synthesis, while the if-needed procedures correspond to goal synthesis, since they replace a set of goals with a structure of goals and subgoals that combine to satisfy the original goals.

One could conceivably avoid the use of if-added procedures entirely, by making all procedures goal-oriented, and "reasoning backward", so that nothing is done until it must be done to meet a specific goal. This runs into difficulty since it allows for little coordination between the goals and the context, so that it is possible to generate vast numbers of irrelevant, impossible, and costly subgoals. Additionally, unless the results of subgoals are added to the knowledge base one can needlessly repeat many subgoal computations. Alternatively, one can "reason forward", taking the contents of the context and applying if-added procedures to derive all the goals which could be met given the context. If this approach is implemented in an unrestrained fashion one can end up swamping the data base in irrelevant results before getting around to meeting the specific goals posed to the system.

The concept of a CKS for a class of goals provides a handle on the problem of how far to let the if-added procedures run. In essence, the if-added procedures become "potential goal" directed, as compared to the "specific goal" directed if-needed procedures. The CKS to be produced by the if-added procedures stands in for the entire set of potential goals. While it may be said that if-added procedures are always directed toward satisfying a set of potential goals, the explicit design of a CKS for a given set of goals provides a mechanism for keeping track of decisions as to what knowledge is to be encoded in the if-added as opposed to the if-needed procedures. Given a set of goals, one needs only to define a class of CKS to organize contextual information and simplify the execution of the corresponding if-needed procedures. If-added procedures then implement a synthesis procedure to build such structures from any of the expected contexts. One can even package such compatible sets of if-needed and if-added procedures into "demon teams" [as in QLISP (Reboh & Sacerdoti, 1973)].

There is a catch to the above suggestion--how does one find a compatible set of if-needed demons whose operations are facilitated by a single reasonable contingent knowledge structure? While we have no general answer to the problem, the technique used in system understanders is suggestive. Essentially, one replaces the search for a CKS and compatible set of if-needed demons with a search for a query language in which all the demons' information needs can be expressed. Starting with a rough idea of those basic semantic entities relevant to the demons (e.g., some "objects", "structures", "relations", etc.), one considers the types of questions about such semantic entities whose answers would help meet the demons's goals. This can often be refined to a well-defined set of primitive questions and composition operations which can be used to answer all of the needed questions. One can then design data structures and procedures that facilitate answering all the questions in the query language, and the data structure and procedures form the class of CKSs.

C. Higher-Level Structures and World Knowledge

The raw input presented to an understander is insufficient to tell the understander all it needs to know about the specific situation it is in. There is always a need for world knowledge in the understander to fill in the meaning of the input.

Several chapters in this book discuss the problems of how to organize such higher-level knowledge, how to find knowledge relevant to a given collection of inputs, and how to use this knowledge to provide an interpretation of the input. We refer below to the process of combining higher-level knowledge with input information as frame-instantiation (after Minsky, 1975; also Kuipers, Chapter 6 and Winograd, Chapter 7) and call the resulting structure an instantiated frame (even though the higher-level knowledge may be a script or scheme, etc.).

The SCA model presented in this chapter does not say much about finding knowledge that is relevant to a given situation, but it does say something about frame-instantiation and the use and structure of instantiated frames. In fact, the process of frame-instantiation is a synthesis operation, and the resulting instantiated frame is a CKS. The primary use of the instantiated frame is to provide the information needed by analysis procedures, in the best organized form.

The view of parsing and perceptual processing as synthesists (and hence cousins to frame-instantiation) leads to the realization that different instantiations of the same frame can be as different in structure as two distinct sentences, or two distinct arrangements of blocks. Many descriptions of frame-instantiation give the impression that all instantiations of a given frame have similar structures, differing primarily in values that fill in slots in the frame. Since slots can be filled by instantiated frames this can indeed lead to a structure as complicated as a parse tree, but not obviously to a network-structured entity like a model for a collection of blocks. Simply expanding slots into subordinate frames is equivalent to top-down parsing. For complicated structures a combination of top-down and bottom-up approaches may be advisable, and one might usefully apply many of the techniques of natural language parsing, including well-formed substring tables or charts for handling local ambiguities until they are resolved by more global constraints.

The use in synthesists of relaxation techniques and other methods for simultaneous constraint satisfaction extends the range of frame-instantiation operations beyond those commonly considered. Much attention has been given to the problem of determining which frame is to be instantiated, and how to switch from attempting to instantiate one frame to the instantiation of another frame when difficulties arise [see the working paper by Fahlman quoted by Minsky (1975)]. This suggests that choosing the correct frame is difficult, but instantiating it is simple. Once one realizes that frame-instantiation may involve complicated simultaneous constraint satisfaction or expansion of structures as complex as natural language parse trees, it is clear that complicated synthesis techniques are needed for frame-instantiation.

The literature on frame-instantiation and the structure of frames gives little idea about the uses to which instantiated frames are put. The SCA model suggests that it is vital to know the questions which are to be answered with the help of the instantiated frame. A synthesist, CKS and analyst are designed together, as interdependent modules, suggesting that both frame-instantiation and the structure of instantiated frames must *take into account the set of operations to be performed on the instantiated frame.* The conceptual structure and machine representation of a CKS depend heavily on the expected inputs to the synthesist, on the design of the synthesist, and *especially on the operations to be performed by the analyst.* Even when the inputs and their real-world meaning are fixed (e.g., where the inputs are always line segments corresponding to the edges of blocks) the information in the CKS must depend on the questions to be asked by the analyst (e.g., do they ever refer to position, volume, etc.).

V. CONCLUSION

The straightforward SCA model presented above is not a complete description of our current concept of the design of an understanding system. A more complete one is hinted at in the section on the LUNAR question-answering system, in which the synthesist is itself viewed as an expert which is broken down according to the SCA model. In general we hold a belief similar to that of Hewitt, in which programs are composed of interacting active procedural elements or ACTORS (Hewitt, 1973). We feel that individual ACTORS should each be organized in terms of the SCA model, with separate synthesis, analysis, and contingent knowledge structures. The synthesist (or analyst or CKS) of a more complicated ACTOR is built up by the activity of a collection of cooperating ACTORS. The crucial issue then becomes the design of the sociology of ACTORS, that is, the communication and control strategies used to organize the efforts of the independent ACTORS. The SCA model itself is a partial (very partial) answer to this organization problem, since it suggests that the ACTORS which compose a complex ACTOR are organized into three separate groups that interact by well-defined means. We believe that another valuable source of ideas for the sociology is the work of Kaplan (1973) on the GSP natural language system, in which the components that make up a parser are organized into consumer and producer modules that interact with one another.

Given the changing economics of machine architecture, in which it is becoming possible to think of machines with hundreds of interconnected microprocessors, the ability to view AI processes in a way which leads to parallel decompositions may be quite useful. Viewing synthesis as a constraint satisfaction operation leads naturally to implementing it by groups of parallel processes which cooperate to find the best structure to match the given set of constraints. We should point out that the economics that we are approaching is not a new one--it is the economics of genetic systems, the economics of constructing a brain. Given the information needed to define one type of neuron and its pattern of interconnections, it is not substantially more difficult to grow millions or billions of copies. Many of the underlying intuitions that led to the SCA model stem from a study of the interactions of neurons in terms of a model for neuron function suggested by Dr. J. Y. Lettvin (personal communication). We hope that it will someday be possible to unify these disparate sets of ideas. Possibly some of the ideas arising from extensions of the SCA model to arrays of interacting SCA modules may be useful as way of viewing the operation of the brain.

ACKNOWLEDGMENTS

This work has evolved over the period of a year during which we have attempted to isolate and refine the ideas developed during the design and construction of the SOPHIE system. We gratefully acknowledge the contributions of many people over that period. We would like to thank Bonnie Nash-Webber for releasing many ideas from the coils of our often convoluted prose. We are also indebted to many people for conversations which clarified our ideas. Martin Kay and Daniel Bobrow at Xerox PARC, William Woods at BBN, and Ira Goldstein at MIT spent many hours having us explain what we really meant to say, making it possible for us to capture some of our more elusive and important ideas on paper. Raymond Reiter of the University of British Columbia has helped us investigate the relation of our ideas to inference procedures and formal logic. Finally, we would like to thank Susan Chase and Cathy Hausmann whose constant encouragement was vital during the months of writing and rewriting necessary to get these ideas on paper. At long last we can finally say to them "the paper is finished" and be sure we mean it.

This research was supported in part by the Army Research Institute--Educational Technology Work Unit.

REFERENCES

Greenfeld, N. Computer system support for data analysis, REL Project Report No. 4, California Institute of Technology, 1972.

Hayes, P. Some problems and non-problems in representation theory. *Proceedings of the A.S.B Summer Conference*, Essex University, 1974, 63-79.

Hewitt, C. Description and theoretical analysis (using schemata) of PLANNER: A language for proving theorems and manipulating models in a robot. Ph.D. Thesis (June 1971) (Reprinted in AI-TR-258 MIT-AI Laboratory, April 1972.)

Hewitt, C., Bishop, P., & Steiger, R. A universal modular ACTOR formalism for artificial intelligence. *Proceedings of the Third International Joint Conference on Artificial Intelligence*, 1973, 235-245.

Kaplan, R. A general syntactic processor in natural language processing. In R. Rustin (Ed.), *Natural language processing*. New York: Algorithmic Press, 1973.

Minsky, M. A framework for representing knowledge. In P. Winston (Ed.), *The psychology of computer vision*. New York: McGraw-Hill, 1975.

Randall, D. L. Formal methods in the foundations of science. Ph.D. Thesis, California Institute of Technology, 1970.

Reboh, R., & Sacerdoti, E. A preliminary QLISP manual. Stanford Research Institute Artificial Intelligence Center, Technical Note 81, August 1973.

Roberts, L. Machine perception of three-dimensional solids. Technical Report 315, MIT Lincoln Laboratory, May 1963.

Sussman, G., & McDermott, D. From PLANNER to CONNIVER - A genetic approach. Fall Joint Computer Conference. Montvale, N. J.: AFIPS Press, 1972.

Thompson, F. Dynamics of information. *The KEY reporter*, winter 1972-73, *38*(2).

Waltz, D. Machine vision, understanding line drawings of scenes with shadows. In P. Winston (Ed.) *The psychology of computer vision*. New York: McGraw-Hill, 1975.

Woods, W. A. Transition network grammars for natural language analysis. *Communications of the Association for Computing Machinery*, 1970, *13*(10), 591-606.

Woods, W. Progress in natural language understanding: An application to lunar geology. *AFIPS Conference Proc.*, 1973, *42*, 441-450.

Winograd, T Understanding natural language. New York: Academic Press, 1972.

Winston, P. Learning structural descriptions from examples. In P. Winston (Ed.), *The psychology of computer vision*. New York: McGraw-Hill, 1975.

SOME PRINCIPLES OF MEMORY SCHEMATA

Daniel G. Bobrow
Xerox Palo Alto Research Center
Palo Alto, California

and

Donald A. Norman
Department of Psychology
University of California, San Diego
La Jolla, California

I. INTRODUCTION

A fundamental aspect of the structure of material contained within a large, intelligent memory system is that the contexts in which units of the stored information are accessed are critically important in determining how that information is interpreted and used. In this book (as well as elsewhere) there are numerous proposals for the representation of information within memory. Most of the schemes currently under active consideration can be viewed as variants of list structures or semantic network structures. All these proposals have a number of common features, including context-independent linkage between units, and separation of processing and data elements. In this chapter we propose a different form for the representation of information which embodies the opposite assumptions about linkage and the separation of data and process. We examine some implications of these memory structures with respect to how the connections among different memory units are formed and interpreted, and we examine some of the issues of processing that arise when these memory structures are used.

The form of our structures is an amalgamation of the principles from the literature on semantic networks, (for example, Norman, Rumelhart, & LNR, 1975; Quillian, 1969) the literature on actors (Hewitt, Bishop, & Steiger, 1974; Kay, 1974) and the new ideas on "frames" (Minsky, 1975; see also Chapters 6 and 7 in this volume). We call our structures *schemata* to emphasize that they differ somewhat from any existing proposals. The word "schema" is taken from the psychological literature, where it has had a long history, most commonly associated with the work on memory by Bartlett (1932), and by Piaget. We propose that one schema refers to another only through use of a description which is dependent on the context of the original reference. We also propose that these schemata are active processing elements which can be activated from higher level purposes and expectations, or from input data which must be accounted for.

II. MEMORY ACCESS USING DESCRIPTIONS

An important property of human memory is the propensity to find analogical or metaphorical references. One event tends to suggest other events. Sometimes the relationships among the two events are, at best, metaphorical. Sometimes, only some limited aspect of the one event is related to the other. The nature of memory retrieval in humans is, of course, not well understood. We have no hard evidence on the paths followed in the effort to retrieve a particular piece of information or of the sorts of events that one is reminded of while experiencing or remembering another. Despite the lack of firm evidence, we think it important to study memory structures that provide these flexible referential properties. Our goal is to specify a memory structure that allows one schema retrieved from memory to suggest others that should also be retrieved, and that is so constituted that it yields analogical and metaphorical retrieval as a fundamental mode of its operation.

In this chapter we speculate on the nature of memory reference processes that can lead automatically, without particular effort, to the richness of the retrievals that we believe to be a fundamental property of human memory. We suggest that memory units refer to one another through the use of descriptions. One memory schema refers to another by describing the other, perhaps by means of a short list of properties of the other. There are different levels of descriptions possible. At the one extreme, a description can be so complete that it unambiguously specifies a unique memory referent. At the other extreme, a description may be so vague that it fits almost every memory referent. We suggest that descriptions are normally formed to be unambiguous within the context in which they were first used. That is, a description defines a memory schema relative to a context. In novel contexts, a description yields novel results. We call such descriptions *context-dependent descriptions*.

A. Context-Dependent Descriptions

A context-dependent description needs only to be sufficiently precise to specify the desired referent with respect to the context in which it is used. A description contains the important properties of the information relative to some context. This reliance on the power of context is perhaps the most important aspect of retrieval through description. It means, in essence, that a retrieval mechanism must use two sources of information in determining the referent that it seeks: the description and the context. The context delineates some restricted set of elements within the memory that are relevant to the situation: we call these elements either the *focus elements* or the *focus schemata*. The description selects a set of possible candidates from the focus elements. In the ideal case, the description selects a unique candidate element from the focus elements which is the referent being sought. In other cases, there may be no candidates or several candidates, and special processing must occur to resolve the difficulty.

Examples of the combined power of partial descriptions and context are readily available from consideration of perceptual phenomena. For instance, cartoon drawings rely heavily on the fact that although the lines and marks on the paper only provide suggestions (or partial descriptions) of the intended objects, the context created by the overall drawing makes the interpretation of those lines and marks possible. The retrieval mechanism must be designed to cope with close mismatches and with multiple matches. Most likely, it should operate by attempting to return a single schema in response to a request, even if several schemata were possible, or even if no single schema satisfied all of the description (as long as the violations were not severe). In using an old schema in a new context, this best-match strategy allows identification of analogic similarity.

Consider an example. Suppose that a particular event is witnessed, say that of a large dog (Spot) in a fight with a smaller, relatively weak one (Rover). Given the contextual setting in which the only objects of note are the two dogs Spot and Rover, the description of the scene could be simply a formalization of the statement that two animate objects are present, and the smaller one attacks the larger. Obviously, this is not a complete description; it relies heavily on the contextual setting. In this context, we may be told to associate the term "underdog" with the object in this situation which has the description "small, animate" (that is, Rover). At some later time, if the original setting is retrieved (which identifies Spot and Rover as the only animals in the scene), then this relatively minimal description uniquely identifies the role of each participant.

Suppose now that a new setting occurs, say one with a small person in a fight with a larger one. It causes the perceiver of the scene between the two people to be reminded of the earlier scene between the two animals.

The schema for the fight from the earlier scene is directly applicable in this context because of the minimal description used to refer to each participant in the schema. Because this schema is used, the term "underdog" is linked to the smaller person by direct association, in what might otherwise be considered an analogic match. To recognize where the association was derived from, the complete earlier setting would also have to be retrieved, and the perceiver would have to recognize that the old description which applies to the new setting was derived from this different setting, one with dogs as the characters instead of people.

A single description can apply in many contexts. Thus the minimal description of the fight between the dogs is useful even in inappropriate contexts. Suppose we had the situation of Don Quixote attacking a windmill, or a single individual making a loud, public attack on a large corporation; it is still desirable for the retrieval mechanism to be sufficiently flexible that it can match the windmill or the corporation with Spot, and to match Rover with the protagonist. To do this match, the retrieval system would have to relax the restriction that the attacked object be animate.

B. The Form of a Description

One fundamental issue in determining the form of a context-dependent description is that of deciding on the terms that can be used within a description. Presumably a description contains as its members other descriptions as well as some constant terms. A constant is a description which retrieves unambiguously a single schema independent of context. Initially, in building up memory, some absolute reference points in the memory structure are necessary; built-in constants or primitive terms can serve this purpose. Primitive terms probably consist of grammatical case relations, a basic set of primitive operations, measuring operations, and dimensional terms for spatial and temporal representation. The sensory systems must each contain their basic dimensional primitives, and there must be primitive terms for the concepts and acts that are known by or that can be performed by the system. Once good higher level bases (descriptive terms) for constructing descriptions have been created, descriptions can consist of only non-primitive terms. The major implication of this possibility is that the "style" of encoding (the choice of terms used) must be reasonably consistent. If not, there will often be the "paraphrase problem" of deciding on the equivalence of two nonidentical descriptions. We believe that such style conflicts are rare within coherent sets of schemata, and that therefore reduction of descriptions to primitive elements (as might be advocated by Schank, Chapter 9) is not necessary, nor even usual. Style conformity is aided by a fundamental mode of forming a description. We believe that many descriptions are formed by identifying the schema sought as an instance of another schema, with the new one further specified, or with some changes or exceptions. The *isa* link of semantic networks, and the *beta* notation of Moore & Newell (1973) are examples of the formation of a description (or a schema) by identifying it with another, with certain specified differences.

C. Properties of Context-Dependent Descriptions

Use of partial description and context for reference provides a number of features which we feel are important in a memory system. These are:

* *Efficiency.* Because context is used as part of the address specification, the descriptions within a schema can be short and efficient, providing only enough information to distinguish the referent in context.

* *Generalizability.* A description makes a schema into a generalized form. The same schema can be used in different contexts without changing the descriptions. In the new context, the descriptions contained within the schema will refer to memory structures which have the same relative properties with respect to the new context as the originally intended memory structures had to the original context. Thus memory access by context-dependent description automatically makes a unique, particular schema

into a generalized schema whenever the context is changed. Metaphorical and analogical use of schemata becomes a direct result of the representational scheme and does not require any special mechanism. (In fact, it requires special mechanisms to prevent analogical and metaphorical extension.)

* *Approximation.* A description used in a novel context allows for close matches to be retrieved. Close matches can focus attention on errors or on significant differences in the context from the original.

* *Reliability.* A context-dependent description allows for graceful degradation of function in case of error either in processing structures or in memory structures. Because descriptions are relative, and because the system is designed to cope with descriptions that yield only partial matches or ambiguous matches, any error that produces these results can be handled smoothly. An error in description or process is treated simply as a case of analogical or metaphorical match, and when a failure to match a description occurs, the system can still return with the best possible match (plus a statement of the mismatching aspects).

* *Currency.* Context-dependent description automatically provides a mechanism for referring to the latest version of information. As long as newly acquired information fits a previously determined description, the new information will be retrieved whenever the old description (and the appropriate context) are invoked. This updating requires no change to either the old information or the old description. Of course this implies that a new item with a description similar to an old one will interfere in the retrieval of the old one (and vice versa).

* *Partial knowledge.* Even if the description and context are insufficient to specify the referent, some knowledge is still available. First, it is apparent in the schema that there is a referent. Second, some aspects of the referent are known and can be used by the system. Finally, a memory procedure can be set in operation to

monitor memory for the appropriate referent. As processing continues and as the information relevant to the development of the contextual setting accumulates, previously uninterpretable referents may suddenly become obvious.

III. PROCESSING STRUCTURES

A data structure that is built around schemata which use context-dependent descriptions has a number of implications for the processing structure within which it is embedded. The retrieval mechanism must be reasonably powerful, for it must combine the information from both description and context to determine the set of possible memory schemata relevant to the situation. In addition we propose that each schema is a self-contained memory structure, capable of performing operations because it contains procedural definitions of its potential functions and operations. In general, we conceive of a large number of these active schemata all operating concurrently in a supportive environment, each drawing computational resources from some central pool, each receiving inquiries and generating messages. At this point, we need to specify the form of this operating environment and the general principles of processing that we believe necessary.

Consider the human information processing system. Sensory data arrive through the sense organs to be processed. Low-level computational structures perform the first stages of analysis and then the results are passed to other processing structures. In the awake human, high-level conceptual activity is normally always in progress. Incoming sensory information is either assimilated into the ongoing cognitive processing or causes an interruption of the ongoing processing. There is a large literature on various aspects of psychological processing relevant to our discussion. Basically, the literature from experimental psychology on attention indicates that the central high-level cognitive mechanisms have a limited processing capacity. When a person is concentrating intently on one demanding activity, that person is able to do very little processing on other activities. Depending on the nature of the tasks, there are rather severe limits on the nature of the activities that may be carried out at the same time. In general, one does not err much by assuming that only a single high-level cognitive task can be performed at any given time. Time sharing of two unrelated tasks, if possible, must be performed with a reasonably slow switching rate, with the time spent on each task between switching measured in seconds.

The limit on the number of activities performed at once does not imply that sensory inputs which are not attended to are ignored, however. Some types of sensory information can be processed, even when the system would appear to be fully loaded. Simple signals can always be detected and complex signals may be detected, although not always recognized. "Important" signals, however, are often capable of attracting attention away from the ongoing activity. The classic examples in the literature are the observations--which have been experimentally confirmed--that although people may be so busy at a task that they claim not to "hear" words or sounds directed at them, they will frequently respond if their name is spoken or if the word fits into the context of the sentence which they are processing at the moment. The problem that these observations pose for theoretical psychology, of course, is that when the importance of a signal is measured by its semantic content, then importance cannot be determined unless the word has been processed. If, however, all words are processed deeply enough to determine their meaning, what does it mean to be so busy performing a task that "nothing" else is attended to? The psychological literature suggests several mechanisms for handling this problem: in this chapter we present a new proposal for this and related problems.

A. Basic Processing Principles

Our analyses of the properties of psychological data and phenomena suggest to us that the human processing system has a number of fundamental principles which underlie its operation, specifically:

* *The processing system can be driven either conceptually or by events.* Conceptually driven processing tends to be top-down, driven by motives and goals, and fitting input to expectations; event driven processing tends to be bottom-up, finding structures in which to embed the input.

* *All the data must be accounted for.* This implies that incoming signals require processing at some level. Thus, a schema to account for a clock's ticking will accept a tick with no further processing demands. If a tick is not heard at the expected time, this is also a datum that must be accounted for.

* *There is a limit to the processing resources available to the organism.* This limit may vary with arousal, but in situations requiring performance on more than one task, each can be allocated only a fraction of the then available resources.

When the resource limit principle is combined with the preceding two, it accounts for interesting aspects of processing behavior. The limited ability to perform several tasks well simultaneously occurs when the resource require-ments of the data-driven tasks exceed the limit on processing capacity. Ongoing processing is interrupted even when the system is heavily loaded with other tasks because all the data must be accounted for. When processing demands exceed the available limit, a deterioration of performance results. The deterioration usually takes place gracefully, however, and not abruptly. This point is elaborated on in the following section, and is treated in depth by Norman & Bobrow (1975).

B. Data-Limited and Resource-Limited Processes

Here we summarize briefly the points made in more detail in the paper by Norman & Bobrow (1975). When two (or more) processes use the same resources at the same time, they may both interfere with one another, neither may interfere with the other, or one may interfere with a second without any interference from the second process to

the first.   The important principles are that a process can
be limited in its performance either by   the amount of
available processing resources (such as memory or processing
effort) or by the quality of the data available to it.
Competition among processes can affect a resource-limited
process, but not a data-limited one.

Consider the problem of performing a complex cognitive
task.   Up to some limit, one expects performance to be
related to the amount of resources (such as psychological
effort) exerted on the task.   If too little of some processing
resource is applied, say because processing resources are
limited by competition from other tasks being performed at
the same time, then one would expect poor performance.
As more resources are applied to the task, presumably
better and better performance will result.   Whenever an
increase in the amount of processing resources can result in
improved performance, we say that the task (or performance
on that task) is *resource-limited.*

Now consider the problem of performing some simple
task, say of identifying a sound which is embedded in
noise: the processing is limited by the quality of the data.
Once all the processing that can be done has been
completed, performance is dependent solely on the quality
of the data.   Increasing the allocation of processing
resources can have no further effect on performance.
Whenever performance is independent of processing
resources, we say that the task is *data-limited.*   In general,
most tasks will be resource-limited up to the point where
all the processing that can be done has been done, and
data-limited from there on.

Operations which share the same limited capacity
mechanism will not interfere with one another until the
total processing resources required by all exceeds some
maximum.   Moreover, in any given range of resource
allocation, one process may interfere with others, but the
others need not interfere with it.   Just what kind of
interference effects are found depends on the particular
form of the performance-resource function for each process.
Interference can only be observed when a process is
operating within its resource-limited region.

Note, therefore, that the effects of interference need not
be symmetrical.   If task A interferes with task B, but not

the reverse, then it would be incorrect to conclude that one of these tasks does not require processing capacity from the same central pool as the other. On the contrary, interference in either direction implies that both tasks draw resource from the same common pool. The asymmetry in effect results when one task is data-limited while the other is resource-limited. Wherever two tasks show an asymmetry in interference effect, it should be possible to demonstrate interfering effects on both by a sufficient change in the availability of processing resources.

One can change the available resources either by increasing or reducing the demands of existing tasks or by adding or removing tasks. Some caution must be used in deciding whether or not one has managed to change resource allocation. If some data-limited task requires some minimum resource $R\text{-}min$ to operate at all and then operates at its best performance, then the only way to change its demand on resources is either to remove it or to add it anew: no partial allocation of effort is possible.

C. Event Driven Schemata

Schemata are event driven. By this, we mean that all input data automatically invoke processing. These input events must be accounted for. Such inputs generate descriptions which are fed to a number of potential contexts of interpretation, some of which may be suggested by the descriptions themselves. If a quick match is found, the sensory input is fit into a context. The context may itself be a nonprimitive sensory construct whose description might allow it to be fit into a higher level context schema. Associated with a schema may be procedural information which indicates an action to be taken if an instance is found. Such action may demand only low-level responses (for example, having seen a desk, be prepared to see a chair), or may request full use of the central processing facilities ("why was my name said?"). Other internal events can also invoke automatic processing. The recognition of a familiar object in unfamiliar surroundings may trigger special actions.

The amount of processing actually done for a request is,

of course, mediated by the total processing load on the system. Schemata that are invoked by sensory events usually cause only low-level decision processes to occur, but the more conceptually based the required decision process, the more processing effort required: here, the resource limitations can severely limit the performance capability.

Consider an example. In driving an automobile while deeply engaged in some other activity--perhaps talking, listening to a conversation, or thinking--the amount of processing effort left over for the driving is much reduced. Driving, however, is a task that automatically creates a continual flow of new sensory signals, and these sensory signals usually demand low-level processing sufficient for the driving to be done; in general we cannot so distract ourselves by an interesting alternative activity that we entirely neglect the relevant driving activity. Although the mechanics of driving can take care of themselves at a low-level (this is true of most over-learned event-driven activities), higher level cognitive aspects of the driving task are not usually event driven, and they will suffer. Thus if too deeply engrossed in other tasks, the overall level of driving will suffer. No planning activity will take place. An impending decision point may not be anticipated, so that braking and steering activity will take place only when the sensory signals require them, not at an early enough time to ensure smooth, high quality performance.

An important feature of our proposed processing strategies is that, although all the data must be accounted for, it does not really matter how. We believe that there is sufficient flexibility in the use of schemata that an incorrect or very general accounting for data does not cause harm. When sensory events are misinterpreted, for most purposes it will not matter, if only because we simply do not care about most sensory events. For most purposes the original interpretation is quite adequate. When better interpretations are needed, then the schemata can be expanded or modified to provide them. Initially, all the data must fit into some schema, but it does not matter if the fit is bad.

D. Depth of Processing

Everything that arrives at the organism must be processed to some extent. Because the processing resources of all devices, including the human, are finite, there must be something that distributes the processing resources that can be allocated to any task: there must be some scheduling device.

What things should be processed in depth? We argue that it is most important to process what is least expected. If an event occurs that is totally expected, then there is little information to be gained from its detailed analysis. If the event deviates from expectations, or if an event that is expected fails to occur, or if an event that one is not prepared for does occur, then these are special events and must be given priority in processing. Thus it is that the things that we most expect to see or experience will leave the least impact on us: it is the discrepancies that we will note. Moreover, the same basic principle tells us how much to process discrepant events: we process them until we know how to account for them. At that point, they are no longer discrepant and, therefore, no longer need processing.

When we say that "all the data must be accounted for", we mean simply that some conceptual schema must be found for which these data are appropriate. If the data are seen not to be of importance to the central analyses of the moment, then almost any schema will do. If the data appear to be important--and this importance is determined by the nature of the schema for which they appear to be relevant--then processing in depth will probably be necessary to elaborate on the manner in which those data are interpreted beyond that provided by the initial schema. Finally, if the data cannot readily be accounted for, then we suspect they create an interruption in the processing cycle, for they will demand sufficient resources from the system to enable them to be processed sufficiently to be understood at whatever level is necessary.

The psychological literature on memory indicates that events that are not processed deeply are not well remembered: the deeper processing, the better the memory for those events (see Craik & Lockhart, 1972). One would expect, therefore, that data which were readily accounted for would not require much processing, would not be well remembered, and probably would not receive any conscious attention. Data which either were deemed to be important or which could not easily be accounted for would, however, receive sufficient processing effort and, as a result, they would probably be remembered later. Moreover, we suspect that they would receive conscious attention at the time of their arrival and processing. Thus data which are expected or otherwise readily accounted for would be ill remembered. This would help explain why we need not be concerned with every detail or anomaly of the environment. To use an example provided by Abelson (personal communication), a red stain (tomato) on the manuscript copy of this chapter could be accounted for by low-level organizational schemata (namely, the schema for stains and shadows). We would not necessarily even be aware of the stain, despite the fact that a reasonable amount of processing effort was expended in accounting for it. It is only if the stain could not readily be accounted for that it would reach conscious awareness (as would be the case, say, if the stain would move about on the page or float one inch above it). Events which are very close to expected events may also be assimilated to their expectations. In this case, the differences will probably not be remembered, only that the general schema was instantiated. The example given by Kuipers (Chapter 6), where with a quick view, a modern clock was assumed to have hands, is typical of human perception.

E. The Organization of Processing

We view the cognitive processing structure as one that consists of a multilayered assemblage of experts. Each expert is a process that knows how to handle the data and suggestions provided it. When situations arise that an expert cannot handle, or when communication with the other experts that it knows about fail, then it passes on its information and messages to higher level processes. The entire system consists of a multiplicity of hierarchies of experts, each expert working on its own aspect of processing, interpreting and predicting the data which are available to it, shipping requests to higher processes, and expectations of inputs to lower ones.

An important aspect of the organizational structure of processing concerns the interactions of conceptually driven and data-driven schemata. If the system is to have any function at all, there must exist several overriding considerations. There must be purposes to activities. There must be some procedure for selecting from among all the various activities taking place at any moment those that are most important for the purpose of the system. Basically, we believe that the system must be provided with motivations to provide top-down drives, a capacity to learn, and the ability to be aware of itself. We conclude that there are reasons to postulate a single central mechanism having many of the properties ascribed to human consciousness.

Purpose. Purposes add direction to the system--the top-down hypothesis driven aspects. The principle that all the data must be accounted for adds the bottom-up drive. Both would seem to be essential. Without purpose, the system will fail to pursue a line of inquiry in any directed fashion. Purposes should be at a high level, not local, simple goals. A high-level purpose coupled with sufficient operating principles should thereby automatically produce the necessary subgoals for the immediate demands of processing, and provide criteria for allocation of resources to event driven schema relevant to the purpose.

Motivation. A person can pursue several purposes at one time. One can, for example, be driving home, trying to find a good music station on the radio, and having a conversation. When the demands from these three tasks exceed the capacity of the processor, criteria from the individual purposes cannot mediate conflicts for demands that have differing purposes. There must be a central motivational process which serves this function for most conflicts.

Retrieval and evaluation. The distribution of computing resources should be guided by the principle that all the data must be accounted for: effort is spent processing data that do not fit into any active schema. The existence of data that do not fit an existing schema, or the absence of important data that are required by a given schema are both capable of requesting some central mechanism to examine the nature of the mismatch. The retrieval mechanism must be capable of the evaluative role that must be performed in assessing context-dependent descriptions. Descriptions must be combined with context, allowing metaphorical or analogical retrieval to take place and to be used to useful purpose. We believe that some central mechanism that has access to many memory schemata is essential in performing intelligent evaluation whenever a memory schema has proposed an unsatisfactory match for a description.

A central mechanism. We believe that all these considerations together require that the system be guided from the top by a single central mechanism, one with awareness of its own processes and of the information sent to it by lower order schemata. We believe this central conscious mechanism controls the process that schedules resources, initiates actions by making decisions among the alternatives presented to it, and selects which conceptualizations to pursue and which to reject. We assume that this mechanism keeps track of its operations and of the overall context by means of a small capacity memory structure, probably the short-term memory structures that are widely discussed in the psychological literature. We believe this central evaluating mechanism is probably serial, probably slow, and probably resource-limited. One major argument for the existence of a single, central control mechanism is that despite the multiplicity of processing structures, there is only one body. There must be coherence and unity in the overall control. Conflicts must be resolved and important decisions must only be made once. These statements do not mean that there cannot be several high-level mechanisms, each specialized for certain types of decision or control functions, each perhaps having different modes of operation. The important point is that at any specific time, for any given task or for any given process, only one of those mechanisms must be in control at any moment.

IV. SUMMARY

In conclusion, we propose that memory structures be comprised of a set of active schemata, each capable of evaluating information passed to it and capable of passing information and requests to other schemata. We suggest that a memory schema refers to others by means of a description that is only precise enough to disambiguate the reference within the context in which the original situation occurred. This context-dependent description thereby provides an automatic process for creating general memory references from specific events, allowing for automatic generation of analogical or metaphorical memory matches. The retrieval mechanism that operates upon the descriptions must be intelligent enough to combine both descriptions and context in a meaningful, useful manner, and it must be relatively insensitive to mismatches and underspecification.

The processing structure of the memory system is one that has a limit on resources that are available Any given process is either data- or resource-limited. Some scheduling device is necessary to keep things operating smoothly. We believe the system to be driven both by the data (in a bottom-up fashion) and conceptually (in a top-down fashion). The principle that "all the data must be accounted for" guides the bottom-up processing. We believe that a single, conscious high-level mechanism guides the conceptual processing, taking into consideration the motivation and purposes of the organism.

Conscious processes are invoked whenever underlying schemata provide information for evaluation, whenever new processes must be invoked or old ones terminated, or whenever the output of one schema must be communicated to others not immediately invoked. Any time that there is a mismatch between data and process or expectations and occurrences, conscious processes are brought in. The automatic, active schemata of memory and perception provide a bottom-up, data driven set of parallel, sub-conscious processes. Conscious processes are guided by high-level hypotheses and plans. Thus consciousness drives the processing system from the top down, in a slow, serial fashion. Both the automatic and the conscious processes must go on together; each requires the other.

ACKNOWLEDGEMENTS

This chapter was written while Norman was a fellow at the Center for Advanced Studies in the Behavioral Sciences, Stanford, California, and we are grateful to the Center for the facilities which they provided. Research support to Norman was provided by grant NS 07454 from the National Institutes of Health.

REFERENCES

Bartlett, F. C. *Remembering: a study in experimental and special psychology.* Cambridge: Cambridge University Press, 1932.

Craik, F. I. M., & Lockhart, R. S. Levels of processing: A framework for memory research. *Journal of Verbal Learning and Verbal Behavior*, 1972, *11*, 671-684.

Hewitt, C., Bishop, P., & Steiger, R. A universal modular ACTOR formalism for artificial intelligence. *Proceedings of the Third International Joint Conference on Artificial Intelligence*, 1973, 235-245.

Kay, A. *SMALLTALK, A communication medium for children of all ages.* Palo Alto, California: Xerox Palo Alto Research Center, Learning Research Group, 1974.

Minsky, M. A framework for representing knowledge. In Winston, P. (Ed.), *The psychology of computer vision.* New York: McGraw-Hill, 1975.

Moore, J, & Newell, A. How can MERLIN understand?. In Gregg (Ed.), *Knowledge and cognition.* Baltimore, Md.: Lawrence Erlbaum Associates, 1973.

Norman, D. A., & Bobrow, D. G. On data-limited and resource-limited processes. *Cognitive Psychology*, 1975, *7*, 44-64.

Norman, D. A., Rumelhart, D. E., & the LNR Research Group. *Explorations in cognition.* San Francisco: Freeman, 1975.

Quillian, M.R. The teachable language comprehender. *Communications of the Association for Computing Machinery*, 1969, *12*, 459-475.

A FRAME FOR FRAMES:

Representing Knowledge for Recognition

Benjamin J. Kuipers
Artificial Intelligence Laboratory
Massachusetts Institute of Technology
Cambridge, Massachusetts

I. INTRODUCTION

How can we represent in a computer program the kind of knowledge people manipulate easily and effectively? One of the significant discoveries of artificial intelligence has been how computationally difficult are the elementary tasks of vision, language, and common sense reasoning which we perform continually in the course of our everyday activities. The techniques used by the artificial intelligence programs of the past decade are simply not powerful enough to approach human performance over any wide range of tasks. New mechanisms have recently been proposed by which the organization of previously accumulated knowledge can assist active perception and understanding. Briefly, the idea is that if there is too little computation time when a problem comes up, do some of the work in advance and keep the computed results available. This in itself is not an astonishing insight, though it does focus our attention on the relationship among immediate perception, understanding, and long-term, real-world knowledge. It obviously should be easier to see something which has previously been seen, and the question becomes how to organize and use such previous experience.

Minsky (1975) proposed a theory of "frames" as a mechanism for representing knowledge in the computer. A frame is a structure which represents knowledge about a very limited domain. A frame produces a description of the object or action in question, starting with an invariant structure common to all cases in its domain, and adding certain features according to particular observations. The resulting description is stated in terms of a limited number of descriptors. A critical point is that the frame, as the unit of represented knowledge, is quite large. Rather than being on the order of a single property or relation attributed to an object, it is on the order of a description of the object with additional information indicating relations with other frames. Minsky's paper has evoked a great deal of discussion and interest in exploring further levels of detail. It presents plausible and provocative examples of the application of frames to different problems in artificial intelligence. Since then, Winograd (Chapter 7), Bobrow & Norman (Chapter 5), Fahlman (1973), Rubin

(1975), and others have begun to distinguish the various theoretical and technical issues often grouped together in discussions of frames.

In discussions of frames, there is a tendency for supporters of the idea to have an intuitively satisfying internal model of the theory which they have great difficulty making precise and communicating to others. It can be difficult to distinguish clearly between the concept of frames and previous ideas, or even to state the concepts precisely enough to evaluate them at all. In this chapter, I attempt to provide an intuitive model which can serve as a foundation for more precise statements. With an intuitive example in mind, I extract some of the properties which are desirable in a frame representation. Next, I attempt to distinguish those issues which are relevant to recognition, that is, the problem of selecting one of a fixed set of alternate interpretations for a collection of observations. I present an example of recognition (in a tiny world) for which actual technical decisions are made. In the last section, I discuss the simplified model of frame-based recognition used by the example, and outline the limits of its applicability.

II. IMPORTANT PROPERTIES OF FRAMES

Some of the important properties of frames as a representation for knowledge are listed below, to be discussed in more detail later.

Description. A frame provides an elaborate structure for creating and maintaining a description. A primitive element of this description may be expanded to a frame when its internal description becomes of interest.

Instantiation. This is the process by which the frame produces a description of the object being examined by substituting observed for predicted values. Features whose real properties have not been observed are represented by default values.

Prediction. The frame's predicted description can be used to guide the collection of observations for instantiation. It also produces the defaults which substitute for unobserved features.

Justification. Different features of the frame description have different amounts of confidence. Some are clear observations, others are choices among few alternatives, and others are default assignments.

Variation. The dimensions and ranges of possible variation of each feature are limited and specified.

Correction. Anomalies may indicate that the current frame is not correct, and that a different point of view is called for. The frame can analyze the anomaly to select a more appropriate replacement.

Perturbation. For small changes in the observer or the observed, perturbation procedures correct the description without complete recomputation.

Transformation. In case of more significant changes, transformation procedures propose frames suitable for the new situation.

A. Scenario

Consider for a moment an intuitive description of how a frame system might work in the everyday vision process. As you are walking through an unfamiliar house, you come to a normal interior-type door, open it, and walk through. At the moment that you open the door, your (entirely reasonable) expectations have already brought a "room" frame to mind. There is no delay in comprehending the fact that you see four walls, floor, and ceiling, since you already "knew" that they would be there, even without having seen them. Indeed, if these expectations had not been fulfilled, and you had been presented with, say, a seashore instead, you would experience a sense of disorientation. You have found a room, however, and your (mostly unconscious) analysis continues. The window on the opposite wall is incorporated into the room description which is forming in your mind, very quickly because you have available a number of prepackaged window descriptions. These descriptions are also frames in their own right, but will only be used as stereotypes unless you direct your attention to them. A bed in the room causes the general "room" frame to be replaced by a more specific "bedroom" frame, in which a dishwasher is no longer a serious possibility. The visual information already collected by the "room" frame, however, is still valid and is incorporated into the description within the bedroom frame.

Your attention passes over a clock near the bed and focuses on the fireplace. The fact of its existence and the superficial properties of fireplaces are recorded in the top-level room frame, but another frame is activated to record the description of the fireplace in detail. That information is extraneous in the room frame, and needs a context of its own. When questioned later, you will be able to answer detailed questions about the fireplace (perhaps noticing a subjective feeling of focussing attention on the fireplace and away from the rest of the room when answering), and you will be unable to say more about the clock than that it was a clock mounted on the wall. Quite possibly you will recall it as having hands in spite of the fact that, being a very modern clock, it had none.

In constructing the description of the room, you would have verified in passing that it was a clock, perhaps by noticing the characteristic hour marks, and then allowed the stereotype description of the clock feature to provide the rest. This kind of self-deception by expectation is a result of the diligence of the frame mechanism attempting to extract a maximally detailed description from a minimal amount of input information. I use an example where the default assignment was incorrect because there is less doubt in such cases that the information was supplied by the frame. In general, of course, such stereotypes are correct, making it uncertain whether the information came from a default description or an actual observation.

B. Description

A frame has a small domain of expertise and contains the knowledge necessary to create a description of an object in that domain. Some knowledge tells how to take a set of observations and create a correspondence between those observations and the descriptive mechanism of the frame. Other knowledge allows the frame to predict some features of the description after observing others. Transformation knowledge maintains the description under small changes of viewpoint, to avoid having to redescribe the scene. We can begin to make a distinction between the knowledge in a frame which is about the object being described (the expected features and the relations holding among them), and that which manipulates the description in response to new observations or changes in viewpoint. The latter kind deals with the relations among descriptions and so could be considered as describing the properties of the domain and not of an individual object.

There is an important point to be made about the relation between the local nature of observations and the global nature of descriptions. The global order imposed on the sensory inputs must be learned: it is not intrinsically present in what is seen. Any theory of representation of knowledge, and of recognition in particular, is trying to explain exactly how we impose the order we have learned through experience onto the extremely varied and disordered sensory inputs we receive. The important point, then, is that any global knowledge contained by a description must have come from the internal representation. It could not come from the observations alone. This helps to explain how prior knowledge is not only helpful, but necessary for understanding and perception.

The description of an object includes a number of features of that object (which Winograd in Chapter 7 calls IMPs, for IMPortant elements) and the relations which hold among those features. The description also specifies a limited set from which those features and relations are chosen. It is reasonable to ask about the stove when thinking of a kitchen, but in an "office" frame the stove is not mentioned, not even to say that there is none. The description may also contain information computed from observations, but which is certainly not in the sensory image--for example, how many people can be served at the dining room table.

C. Instantiation

Instantiation is the process by which a frame creates a description from observations of an object in its domain. Part of a frame is a description schema which makes building a description a matter of making a number of simple decisions and choosing from among limited sets of alternatives. Most of the choices involved in constructing a description have already been made by selecting that frame. For example, we know that virtually all rooms are bounded by plane polygonal surfaces, and that almost all of those consist of six rectangles: four walls, floor, and ceiling. Thus the part of the descriptive process that describes the walls can use a quick, simple test for large deviations from the expected four-wall description. If no deviations are noticed, the complex description of four perpendicular rectangular walls can be used in the particular room description. This process, based on our experience with typical rooms and the appearance of typical room-edges from the usual perspective, makes it possible to verify a complex portion of a description in much less time than would be required to generate it from scratch.

Our actual experience with rooms comes mostly from particular kinds of rooms: rooms in homes, offices, schools, and other buildings. As we instantiate the general "room" frame we record characteristics which could belong to any kind of room. At the same time, however, the features we see specify which particular kind of room is before us, and bring in the frame corresponding to that kind of room. This is the process of refinement: within a frame of common characteristics, making decisions which determine a particular and more specialized frame in which to continue the description. For example, in the scenario, upon noticing the bed, the room frame becomes a bedroom frame, which affects some (not all) of the remaining alternatives in the description.

D. Correction

In most common cases of recognition the identity of the object being described is not initially known, so selecting the proper frame to instantiate is part of the problem. The current "best guess" frame attempts to create a correspondence between what it expects to see and the observations actually available. If it runs into an observation which is incompatible with its domain, that observation can often indicate a good replacement frame. For example, in attempting to recognize a large, four-legged animal, a reasonable guess would be that it is a horse. Small horns, however, are incompatible with a "horse" hypothesis, but strongly suggest a cow. A single large horn would suggest a unicorn. Notice, however, that much of the previously gathered information, such as color and location of various body parts, is valid in any of the three potential frames, and need not be observed anew within the new frame. Fahlman (1973) is carrying on research along these lines, and I discuss these issues in more detail in Section III.E.

E. Default Values

When some feature of a description has not been observed, either because it is hidden or because it simply has not yet been attended to, the frame can still make quite an accurate prediction about that feature. This is true even if the object has not been observed at all yet, and the only basis for prediction is personal, idiosyncratic experience. For example, if I mention a beachball, I immediately conjure up an image of a particular ball with red and white stripes. These default values are very weakly bound features of the description in my "beachball" frame. It would take very little sensory evidence to make me replace them with better data for a particular description. On the other hand, if I see a line drawing of a cube, I have a very strong expectation of a hidden corner and three more faces, and these expectations would be quite hard to replace.

These default values have two quite important uses.

The first is in guiding the process of recognizing and instantiating a particular description by suggesting what features to look for and where to expect them. The second is to provide answers to questions for which observations have not yet been made. In this way, the frame represents our inductive knowledge of the world as gained by previous experience with that domain of objects. This use of default values also allows a frame representation to satisfy the "principle of continually available output" (Norman & Bobrow, 1975), which says that a process should be able to provide a result even when its analysis has not yet been completed. A lack of data or processing resources should produce a graceful degradation of the quality of the output, but not prevent results from being produced at all.

F. Variation

A frame represents a certain limited domain, and hence a range of variation for objects which belong to that domain. As we saw in the room scenario, the features of a frame may be frames in their own right, embodying ranges of variation. On entering a room, you are prepared for certain typical pieces of furniture. A park bench or a diamond-encrusted throne would be outside the permissible range of variation in this frame. Such an anomaly may indicate to the correction mechanism that another frame is called for. When a number of features are near the extremes of their ranges of variation, their collective unlikeliness can cast doubt on the applicability of this frame and initiate a search for further evidence which may result in a new frame being selected. This is particularly clear in medical diagnosis, where a set of symptoms may be possible within the frame for disease X, but so unlikely that the doctor orders further tests to search for a more plausible hypothesis.

G. Perturbation and Prediction

There are a number of different circumstances when a frame may be transformed or replaced by a different one.

While sitting in a room, if I turn my head, I bring a previously invisible region into my field of vision and lose a region from the other side, or I may move, changing the vantage point from which I view certain features. These are relatively small changes which cause perturbations of the frame and the description it produces. I may experience larger changes by walking into another room, requiring a prediction of what frame I may need next and repeating the instantiation process. These phenomena are not isolated, but lie on a spectrum which includes looking from outside the doorway, or lying on the floor and looking up. These intermediate cases include more common information from the original frame than leaving the room entirely, and involve a more drastic change to the frame than a perturbation.

The common element to extract from these transformations is the idea of partially changing a description while saving those portions which are still valid for the new version. A transformation in viewpoint does not take place spontaneously. It occurs as the result of some action (perhaps mental) with which we may be quite familiar; familiar enough, in fact, to be represented by a scenario frame. An action, like an object, has a description, which often takes the form of a scenario. Frames may certainly be used to represent the kinds of variability scenarios are subject to, just as vision frames represent variability in visual descriptions. Part of the frame for a given action will be a prediction of its effect on commonly associated objects and environments. When I am walking, the "walking" frame will predict the change in the visual geometry of the enclosing room. Conversely, strong visual cues can be used very effectively, in movies for example, to evoke the sensation of motion.

When the action in question forces most of the description to be redone, as when I walk from room to room, then the transformation consists mostly of proposing possible new frames. In a familiar house I may be able to summon up a fairly complete and accurate description from memory, but in an unfamiliar house I need time to get my bearings. For small perturbations, however, such as moving slightly within the same room, the visual geometry of the outlines of the room may change slightly, but most of the features will remain the same, and appear in corresponding places on the walls. If a piece of furniture looks substantially different from the new angle, its own frame may require a transformation.

Occlusion of objects in the background by those in the foreground can be explained by their relative positions within the room description. I do not believe, however, that people accurately predict such occlusions from their mental descriptions. On looking at a scene, the description I generate is not of the picture I see, but of what I think that scene actually is. I come to conclusions about the global nature of the scene from evidence I have, and fill in with default values where I missed actual perceptions.

H. Extreme Anomalies

An extremely unexpected observation, such as opening that door and finding myself at the seashore, is treated in a more serious way. My dumbfoundedness resulting from this occurrence is not only due to the time it takes to find a "seashore" frame, but I am also faced with evidence suggesting that previously accurate notions of continuity no longer hold. I do have some knowledge of geography, and I am filled with curiosity about how I was transported to the sea without noticing. I may decide to reject the evidence and the attack on continuity by concluding that I am dreaming or have gone crazy. Alternatively, I may retreat back through the door and lock it, or in the best Kuhnian tradition, postpone dealing with such questions while I explore and gather more observations. The point of all this is that an extremely unexpected occurence calls into question not only the predicted frames that have proved to be inaccurate, but also that knowledge which led the prediction process so seriously astray. Such experiences are saved and incorporated into newer versions of the faulty frames when structural revisions become possible.

III. AN EXAMPLE: BLOCKS WORLD RECOGNITION

Frames, then, have an intuitive appeal as a metaphor to explain how people organize and represent their knowledge. An obvious question is, of course, whether this idea is of any help to us in representing such knowledge in computer programs. The next example solves a very easy problem, one for which the machinery developed is quite superfluous. The hope, however, is that the way such problems are solved will provide valuable techniques to be used in solving larger, more realistic problems.

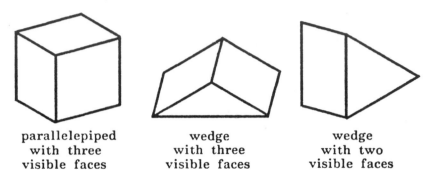

parallelepiped	wedge	wedge
with three	with three	with two
visible faces	visible faces	visible faces

Fig. 1. The domain.

The domain shown in Fig. 1 consists of line drawings of a single, unoccluded block, which can be either a parallelepiped with three visible faces, a wedge with three visible faces, or a wedge with two faces visible. The blocks world has been used as a domain by a number of researchers in different contexts (Winston, 1975; Winograd, 1972), and is rightly criticized as a "toy" world, lacking many of the important and complex problems found in the real world. Much of the difficulty of real world domains comes from our inability to express in a computer program descriptions and distinctions which are obvious (though hard to verbalize) to a human being. The blocks world, however, has very clear descriptive mechanisms, and it is easy to find precise distinctions between two line drawings. In this domain we can focus on the nature of the recognition process, and how the use of frames in manipulating and representing descriptions can aid that process. The hope is that a simple "toy" example will clarify phenomena which would be obscured by other important (but separate) issues in a more complex domain.

Five of the phenomena mentioned in the previous section will be addressed by the example of the block recognition program: description, instantiation, prediction, correction, and transformation. The recognizer instantiates a description of the object it recognizes, using its predictions to guide the recognition. When a conflict occurs between prediction and data, a complaint department associated with the frame selects an appropriate course of action, often a transformation to a new frame.

What, then, does the recognizer take as its input and produce as its output? The "sensory" world of this system consists of a body of data about the line drawing which can be interrogated by asking it questions which are very local, in the sense that a particular part of the visual scene can be reached only by searching along a known edge from a vertex which has already been observed. An attractive metaphor is that of walking over a snowy field attempting to interpret a line drawing laid in pipes hidden under the snow. More precisely, the sensory world consists of edges and vertices, which can perform the following operations upon receipt of the appropriate message.

> A vertex will deliver its type, the edges which terminate at it, and the sizes of the angles between pairs of edges. This corresponds to the result of a "circular search" in the neighborhood of a vertex. The type of a vertex is L, fork, or arrow. The size of an angle can be described as either acute, right, or obtuse.

> An edge will deliver its "other vertex" upon being presented with one vertex. This corresponds to scanning an edge from one vertex to the other.

With this limited sensory world, and even more impoverished descriptive system, the recognizer will attempt to recognize what it sees and provide a global description

of that object. It is important to recognize the difference
between the sensory world which is available, and the
descriptive mechanism which creates an internal
representation to be remembered. Even if the sensory
world provided precise angle measurements, the recognizer
could only describe them as acute, right, or obtuse.
Similarly, people discard or blur many distinctions which
are physiologically available to their senses.

What is the description of a line drawing? A
description imposes a level of organization on the
observational data which is not locally apparent in the
scene itself. Simply by stating that an object is, say, a
cube, the description asserts that a certain collection of
features appears in the scene and that many others do not,
a fact which could be determined directly only by
exhaustively searching the scene. The description also
provides a global structure to the features which is not
apparent in the local relations of the scene. Thus, looking
at one corner of a cube, one may ask of the description,
"Where is the opposite corner?" The scene cannot answer
such a question, for it cannot define "opposite" in a way
that is meaningful to the cube. A third function of the
description is to include properties of the object which are
inferred from the observed features along with the
knowledge of its identity, such as the volume of a cube or
wedge. The description produced by this recognizer will
fill only the first two functions, noting collections of
features and providing a global relational structure. A line
drawing will be classified according to type, and its parts
will be accessible according to the global structure of the
object it represents.

The recognition problem in this blocks world domain is
to select and instantiate the correct frame for the drawing.
Since, however, instantiation must begin before selection
can take place, the recognizer must also evaluate observed
evidence, predict subsequent observations, select a new
frame when necessary, and save previously collected
observations.

Having defined the problem, we can now begin to look
at what the recognizer is. The recognizer consists of three
frames, one for each type of object in the domain. Each
frame is a program for examining the input data and

constructing a description of its type of block from that data. A frame has many of the properties of a description, in that it imposes its own global organization on the observed data and makes predictions based on its observations along with its assumptions about the type of object being observed. An important similarity between a frame and the corresponding description is that a frame will be able to answer questions about as yet unobserved portions of the scene based on its predictions. Thus a frame functions as a complete (though possibly erroneous) description even before its processing is complete.

A frame, however, has additional capabilities which are not present (or necessary) in a description. It contains strategy knowledge which can advise it on the best observations to consider as it builds its description. It also has the ability to evaluate the observations for consistency with the description it is attempting to instantiate, and to turn the process over to a more appropriate frame when a fatal inconsistency appears. During the recognition process this description serves as a hypothesis about the scene which the frame is attempting to confirm or refute. When the hypothesis is refuted, however, it is not only the description which is replaced by a better alternative. The new frame also contains new knowledge about strategy, evaluation, and the handling of inconsistencies in ways that are more appropriate to the new hypothesis.

There are two distinct kinds of knowledge about the features of these line drawings which are embedded in the frame and which guide the construction of the description. The first is local knowledge about the types of vertices which appear in the figure, and how each vertex is connected to its immediate neighbor. The second is knowledge of the global relations which hold among the angles in different parts of the drawing (see Fig. 3, p. 169). These global relations allow an observed angle measurement in one part of the drawing to predict an observation in another part. Both kinds of knowledge serve the same role of predicting observations and guiding the recognition process, but they interact with observations in different ways, and the details of their representation in the frame are somewhat different.

A. Recognition Scenario

Let us follow a scenario of the recognition of a block drawing, in this case the three-face view of a wedge. Fig. 2 shows the stages of the recognition process, with observed data indicated in solid lines and hypothetical knowledge in dotted lines. The first drawing is the actual scene, with the vertices numbered in the order in which they will be explored.

Vertex 1: We start the recognition process by giving the program an initial vertex, which in this case happens to be an L-vertex. The initial hypothesis is that the figure is a parallelepiped, indicated by the dotted lines in the figure. The single angle measurement, along with the parallelepiped hypothesis, predicts the sizes of the four additional angles indicated.

Vertex 2: The second vertex observed agrees completely with the hypothesis, which expected an arrow vertex and had a particular measurement anticipated for the left side angle of the arrow. The two other angle measurements provided by the arrow allow the frame to predict every angle expected in the parallelepiped. Fig. 3 shows the global angle relations which support this extensive prediction.

Vertex 3: This is an arrow vertex, which is the vertex-type predicted by the current hypothesis. At this point we can see that the angle is too small, and that the figure cannot be a parallelepiped. If the program had been given better angle resolution, the angle specialist would also have noticed the error in angle and would have complained to the frame. We are assuming, however that the system cannot discriminate well enough, so the angle specialist accepts the information as consistent, and the recognizer continues with a mistaken hypothesis.

Vertex 4: The fork-vertex at the center of the figure also corresponds completely with the parallelepiped hypothesis. One complete face has now been observed and confirmed.

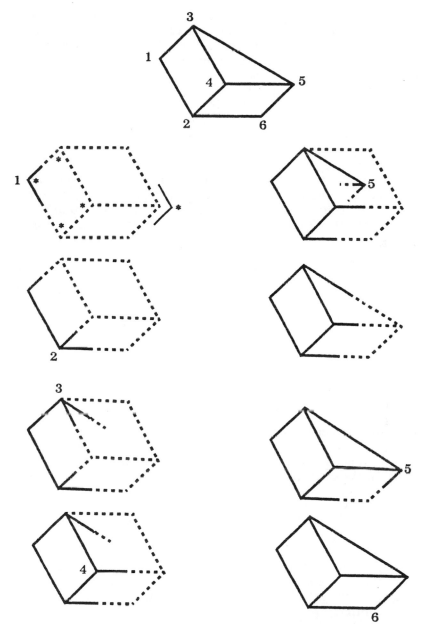

Fig. 2. Stages of the recognition scenario.

Vertex 5: With this observation, the parallelepiped hypothesis finally breaks down. The L-vertex specialist observes an unexpected type of vertex and complains to the frame: "I expected an L, but got an arrow." The parallelepiped frame knows that this particular problem indicates a transition to the three-face view of the wedge. It then analyzes the complaining vertex and the data already collected to discover the correspondence between the cube and wedge frames which will allow previously collected data to be retained. Finally, it executes the selected transformation.

Notice some fancy stepping here. The unexpected arrow vertex was an anomaly to the parallelepiped frame, and the information contained in it could not be completely processed by the L-vertex specialist. Thus it was ignored, and the transition to the wedge frame took place with only the previously known data. Once the new frame was in control, it could deal with the arrow vertex. The arrow vertex, in effect, caused the recognizer to do a "double take".

Vertex 6: At this point, with the three-face wedge frame directing the exploration, there is only one remaining vertex, and it completely confirms this hypothesis. The frame is now fully instantiated.

B. Representation

A frame is built around a hypothetical description. The elements of that description are represented by active program objects (called "specialists") which interact by sending messages to each other. Each vertex in the drawing is represented by a specialist in one of the vertex types: L, fork, or arrow. The properties of that type of vertex are represented by the particular behavior of that specialist. A vertex specialist has pointers to each of the edges terminating at it. An edge is also represented by a specialist with pointers to its two vertices. This network of specialists connected with pointers represents the topological connectivity of the line drawing. The network makes implicit predictions by stating that if a vertex specialist is satisfied with the real (observed) vertex corresponding to it, then a scan along one of the edges should encounter another real vertex which will satisfy that corresponding vertex specialist. Once an initial correspondence has been established between observation and hypothesis, this constitutes a prediction of all the vertex types and their connections throughout the figure. This prediction is embedded in the structure of the frame, and cannot be changed by incoming data, except by refuting the hypothesis and replacing the frame with another one.

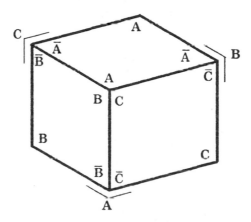

Fig. 3.   Global angle relations in the parallelepiped frame.

New angle predictions, on the other hand, can be freely sent among angle specialists throughout the figure. The relations among the angles in the line drawing (Fig. 3) are represented by angle and relation specialists, who communicate predictions and observations among themselves. By this communication, an angle observation in one part of the figure can affect the prediction in a remote part of the figure. The edges, the faces, and the block as a whole are also represented by specialists, sending messages to each other, whose behavior directs the recognition and instantiation process.

C. The Basic Loop

The basic operation of the recognition process is to select an observation and evaluate it with respect to the predictions made by the current frame hypothesis. The flow of control described here includes the decisions about search strategy, sending observations to corresponding specialists for evaluation, and communicating predictions and additional data between specialists. It is important to notice that these design decisions can be changed independently. For example, the selection of the next observation can be made in a different way without changing the rest of the flow of control. The range of flexibility of these design decisions will be the topic of the last section of this chapter. Since the frame consists of a number of specialists, each with its own behavior, the description of the normal flow of control will also describe much of the behavior of those specialists.

(1) When instantiation begins, an initial observed vertex is sent to the recognizer. Since the initial "cube" hypothesis is symmetrical, the correspondence between hypothesis and data is set up by sending the observed vertex to an arbitrary vertex specialist of the same type. After this, the specialist for the entire block directs the instantiation.

(2) When the block specialist is told to select an observation, it cycles through its faces, telling each in turn to select the observation.

(3) When a face specialist is told to select an observation, it cycles through its edges, telling each in turn to select an observation. If they all refuse, the face passes the refusal back to the block specialist.

(4) When an edge specialist is told to select an observation, it checks to see if it is in a very particular state. It can make an observation only if: a real edge has been observed corresponding to it *and* exactly one of the vertex specialists at its ends has observed a corresponding real vertex. If this state of affairs obtains, the edge specialist performs the scan from one end of the real edge to the other, and sends the newly observed real vertex to its corresponding vertex specialist; otherwise, a refusal goes back to the face specialist.

(5) When a vertex specialist receives an observed vertex, it evaluates the observation against its prediction, by checking to see if the observed type is the same as what it expected. If not, a complaint goes to the complaint department (more on this in the next section). If the type is acceptable, the vertex specialist obtains the real edges and angle measurements which are available from the observed vertex. It sends the observed edges to the corresponding edge specialists, and the observed angle measurements to the angle specialists.

(6) When an edge specialist receives an observed edge from one of its neighboring vertex specialists, it remembers the real edge, and the real vertex at one end, so it can respond differently to future requests for observations.

(7) When an angle specialist receives an observed angle measurement, it compares the measurement against any prediction it might have. A conflict, of course, results in a complaint sent to the complaint department. If there was no previous prediction, the measurement will be of interest to the specialist (called a "relation") which represents the relation among some collection of angles in the figure, so the observed measurement is sent on. An example of such a relation is that holding between the four angles of a parallelogram.

(8) When a relation receives such measurement, it decides whether this measurement implies some useful prediction. If so, it sends that prediction to the appropriate angle specialists.

(9) When an angle specialist receives such a prediction, it simply remembers it for comparison with future observations.

D. The Complaint Department

A frame has a complaint department which receives complaints about violated expectations from the vertex and angle specialists. The offended specialist sends a description of the problem from its own local point of view, and the complaint department, with its more global knowledge, must select the proper course of action. In this example, only the parallelepiped frame has a nontrivial complaint department. There are three distinct responses it can make. It can decide that the observed anomaly indicates that the object being recognized is actually the three-face view of the wedge, for example, and that it can determine the correspondence between what has already been observed and the data expected by the three-face wedge frame. The same can happen to indicate a transition to the two-face view of the wedge. The third alternative (Fig. 4) is somewhat more interesting. The complaint department has enough information to decide conclusively that the new frame should be the three-face wedge, but it does not have sufficient data to select the correspondence between the old and the new frames. It cannot decide which face will be the triangle. The solution adopted in this recognizer is to continue the recognition process under the old hypothesis (now known to be mistaken), under the assumption that the next complaint will be able to settle the question. This decision is based on knowledge of the domain which assures the recognizer that no important data will be lost while working under this mistaken hypothesis. I do not address the question of how such knowledge can be automatically acquired from experience.

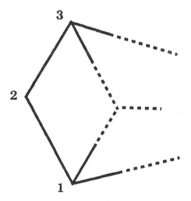

Fig. 4. The ambiguous transition: the frame cannot predict which side will be the triangle.

Table I summarizes the process by which the complaint department deals with anomalies and selects the new frame. The alternative "continue (three-face wedge)" is the case discussed immediately above. The complexity of the complaint department is a result of the number of complaints which are meaningful, and hence of the number of alternate frames known to this frame. This table does not show the fairly elaborate decision procedure for determining the correspondence between the two frames before the transition can actually be executed.

TABLE I.
The Complaint Department

Vertex specialist
       expected arrow, got L ==> two-face wedge
       expected L, got arrow ==> three-face wedge

Angle larger than expected ==> three-face wedge

Angle smaller than expected:
  in L vertex ==>   two-face wedge
  in arrow vertex:
    full angle ==> continue (three-face wedge)
    side angle:
      observed L vertex in that face ==> two-face wedge
      else ==> continue (three-face wedge)
  in fork vertex ==> three-face wedge

E. The Transition

Once an anomaly has refuted the parallelepiped hypothesis, and a more appropriate wedge frame has been selected, the problem remains of actually performing the transition. The simplest solution would be to start over, ignoring previously collected data except to indicate a different frame to start with. This form of recognition is a blind, back-tracking search through a space of line drawings. One goal of this example, however, is to show

how frame-based recognition can exploit the similarities between different line drawings to preserve observations collected under a mistaken hypothesis. At the very least, the actual observations of edges and vertices can be mapped from the old description to the new one because the definitions of adjacency and connectivity are shared by all frames in this domain. In favorable circumstances, higher-level descriptive objects, such as a parallelogram face, will remain valid in the new frame without disturbing their internal structure.

The transition from the parallelepiped frame to the three-face view of the wedge has these favorable properties. The differences between the two descriptions are confined to changing one parallelogram face to a triangular face, and adjusting the angle predictions. To accomplish this transition, the parallelepiped frame replaces the collection of specialists which represent one parallelogram face with another collection for a triangle. It transfers whatever data has already been observed to the corresponding new specialists, notifies all concerned neighbors of the change, and the displaced parts of the old description disappear. The internal structure of the neighboring faces changes only in accepting a new pointer. The angle predictions also vanish, but new predictions are solicited from the angle specialists.

The transition to the two-face wedge is quite different, however. The change here involves much more extensive changes to the structure of the description. Just as in the other transition, there is a correspondence between the representing specialists in the two frames, but in this case specialists who correspond may not have the same behavior. Faces which had two neighbors now have only one; vertices which expected to be arrows will now be Ls; and as before, the angle predictions become obsolete. In this case, all that can be salvaged from the old frame are the actual observations, including the connections between them. These observations are transferred to corresponding specialists in the two-face wedge frame, which incorporates its own higher-level descriptive structure. There is still an important saving in observations to be investigated, but not as much program structure can be shared between the two-face wedge and the parallelepiped as was possible between the parallelepiped and the three-face wedge.

F. The Implementation

This example was first programmed and hand-simulated in ACTORS (Smith & Hewitt, 1974). As the ideas continued to evolve, a working implementation in SMALLTALK (Kay, 1974) was written and debugged in less than two weeks. The ease with which the concept could be translated into a working program is primarily due to the novel semantics of these two languages. Both ACTORS and SMALLTALK evolved from the ideas in SIMULA (Dahl & Hoare, 1972), and are what might be called actor languages, as opposed to function or procedure languages like LISP or ALGOL. An actor is a procedure which can maintain an internal state between invocations. Actors communicate by sending messages to each other, and are not constrained to send messages (or control) only up or down a function-call hierarchy. Allowing an actor to maintain an internal state makes it possible for the variables which are intuitively associated with a conceptual object to be associated directly with the corresponding program object.

A certain amount of confusion is possible between the different types of instantiation in this example. A specialist representing a feature of the line drawing (for example, an arrow-vertex specialist) is written as an actor which maintains a certain amount of internal state, and has a certain behavior in response to particular messages. The parallelepiped frame contains three copies of the arrow-vertex specialist, each of which is an instance of the actor mentioned above. These three instances are not identical, but can be distinguished by which other specialists they have as neighbors. The parallelepiped frame, then, is a program which consists of several parts, some of which share program text but have different internal states. This frame is then provided with a source of observational data. Instantiation of the description is the process by which the various parts of the frame establish a correspondence with observational data. To add further to the confusion, we can imagine a scene containing two unoccluded blocks, for which we make two copies (instances) of the entire recognizer, so that separate frames can be instantiated, resulting in two independent descriptions. This third case seems to have no theoretical interest.

IV. WHAT DOES THIS ALL MEAN?

Let us step back now and see what significance this example has in the larger enterprise of representing knowledge for recognition. The overall structure of the recognizer has some applicability to other domains in which greater expressive power is required of the descriptive mechanism. In the following sections, I discuss the general conclusions which can be drawn about the descriptive mechanisms used, and about the interacting modules which supervise the recognition and instantiation processes. Other domains which have been investigated in some depth, and from which I draw examples, are medical diagnosis (Rubin, 1975), and electronic circuits (Sussman, 1973). These other domains can show features which fit into the framework I have developed, but which do not appear in the blocks world. Where possible, I point out the range of applicability of this framework for recognition, and give examples where it does not apply.

A. Representing the Hypothesis

The block recognizer uses three methods to represent hypotheses about line drawings. They are:

   * the vertex-specialists, which know about a particular type of vertex to which they expect to correspond;

   * the network of neighbor pointers, which links the edge- and vertex-specialists, and homomorphically represents the connectivity of the edges and vertices in the drawing;

   * the angle specialists, which represent the global relations among the angle measurements, and actively communicate predictions about particular angles.

This division of representational effort works in the blocks domain because a clear distinction can be made between the different properties to be represented. There are strictly local features (the vertex types), fixed global relations (the connectivity between vertices and edges), and predictive global relations (the angle relations).

Certain other domains fit into the same descriptive framework so that this distinction between local and global features can be clearly made. A good example of this is the domain of electronic circuits, where the connectivity and local properties of components must also be represented, and global relationships among current and voltage measurements at different points can be predicted. A less geometric example with the same logical structure might be representing the time course of certain diseases, where local specialists are able to recognize particular symptoms, the network of connections is a partial time ordering of events, and the global relations may be among the different measurements of a varying quantity, such as blood pressure or white blood cell count.

There are, of course, many domains where the representational structure described in this example does not clearly apply. This is particularly true when features are not as discretely separable as they are in the blocks world. For example, in medicine it can be important to describe the onset of a certain symptom as "insidious", or otherwise specify an indefinite time interval which can overlap with other events. Notice that we are not simply specifying an interval whose endpoints are discrete (though currently unknown), but rather an interval which fails to possess definite endpoints. The network representation described above lacks the expressive power to deal adequately with this phenomenon.

B. Manipulating the Hypotheses

In the previous section, we saw what kind of expressive power is available for representing hypotheses to this kind of recognizer. Now let us consider the structure provided to manipulate those hypotheses. It consists of four parts:

* a module to select the next observation to consider;

* a module to evaluate the observation, comparing it with what was predicted;

* a module to serve as a complaint department, deciding what to do in response to an observed anomaly;

* a module to perform the transition to a new frame, preserving as much as possible of the old information.

These "modules" do not correspond to segregated pieces of program in the block recognizer, but are design units whose implementation is likely to be distributed among the specialists which comprise the frame.
   In the following sections, we will examine these modules individually and see what range of behavior can be expected of them. The important questions to ask of each one are: What is it asked to do? What knowledge can it consider? What answers can it give? This modularized view of the recognition process also has its limitations, again because of the discrete structure of frames linked by explicit transitions. This simplified view of recognition is based on the assumption that recognition proceeds by adopting a single "best guess" hypothesis, and modifying it to a better one in response to an unexpected observation. There is no provision for entertaining several different hypotheses at once, or for leaping to an unrelated frame where no explicit link exists. There are also important questions about sharing knowledge among distinct frames which are not addressed in this domain.

C. Selecting the Next Observation

This module decides which potential observation would be most useful at each point in the recognition process. Once it has selected one, it sends the observed data to the appropriate specialist to begin the evaluation. The interesting thing about this module is the range of information it can consider, and where it obtains that information. In the block recognition example a particular set of considerations is designed into the selector, so it does not answer the questions below each time it makes an observation. Doctors, on the other hand, are trained to ask these questions explicitly in the course of a medical examination.

* Given what has already been observed, which alternate hypotheses are the most likely? (i.e. for differential diagnosis)

* The frame uses observed data to refine its predictions and the description it is producing. Which observations would be most productive at this time?

* The pragmatic context of this recognition act makes certain parts of the description more useful than others. Which are these?

* What costs (e.g. pain, risk, money, doctor's time) are associated with potential observations?

There are, of course, some cases where the relative importance of these factors may be decided once and for all and designed into the selection procedure, and others where the situation must be actively and frequently reevaluated. Differential diagnosis information may be requested by the complaint department in cases where an anomaly has been observed, but a unique replacement hypothesis cannot be selected.

D. Evaluating the Observation

This evaluation is a point of close contact between the representation and the manipulation of the hypothesis. The frame checks an observation against its hypothesis, asking whether that observation is consistent with the predicted description. The discussion of representation above illustrates the local nature of this evaluation in the block recognizer. The appropriate vertex and angle specialists each check the consistency of the new information with their expectations. The complexity lies in the range of potential results of this evaluation. In the block recognizer, only the first two of the following possible reactions can occur.

* The observation is consistent with the hypothesis, perhaps providing additional information to be absorbed by the frame.

* It is inconsistent, refuting the hypothesis, and the specialist sends a description of the problem to the complaint department.

* It is consistent with the current hypothesis, but singles out a special case about which more knowledge is available.

* It, in isolation, is consistent with the current hypothesis, though near the edge of the range of variation. However, enough other observations are also near the edges of their ranges of variation that the frame becomes suspicious and complains to the complaint department.

The third, or "further specification" link between frames provides additional information which allows more detailed predictions or better selection of observations. The fourth possibility allows suspicion to be cast on a hypothesis as it becomes more and more unlikely, even though it may never be conclusively refuted. It may nonetheless be replaced by a better alternative.

E. Selecting a New Hypothesis

This module, the complaint department, is given a description of the current complaint (and perhaps remembers past ones), and is asked to select a new hypothesis. In the block recognizer, most of the possible anomalies simply specify unambiguously the frame which should replace the current one. As we saw above, however, there are cases in which further information is necessary to select the correct orientation for the new frame. In either mode the complaint department must possess knowledge about which alternate hypotheses are available. In most cases of practical recognition these decisions will be reduced to simple tests of the observations, just as in the block recognizer, rather than active problem-solving during the recognition process. The speed of frame-based recognition depends on the assumption that the number of potential alternatives in a domain is manageable, and that most anomalies clearly suggest alternate hypotheses. The eventual answer provided by the complaint department should be a new frame to replace the complaining one. Some of the potential courses of action leading to this result are:

* the anomaly may simply specify a new hypothesis to replace the old;

* there may be previously collected information which can be reexamined in more detail to decide between potential new frames;

* the complaint department may request a particular observation for differential diagnosis from the module which selects the observations;

* if the anomaly is minor, or there are no good alternatives, the current frame may just remember the problem and continue under the old hypothesis, hoping that further observation will illuminate the situation.

The complaint department is also involved in representing the frame's range of variation. Each feature of the frame description has its own range of variation which it will accept before complaining. The complaint department may then decide to disregard certain complaints or accept excuses under some circumstances. A frame system could believe that all dogs have tails, but admit the possibility that a dog without a tail could still be a dog.

F. Translating Knowledge to the New Hypothesis

At this early, somewhat speculative stage of research, it is considerably harder to generalize about the transition procedures than it is to talk about the other parts of the recognizer. The other parts of the recognition process depend largely on the properties of the domain; the transition depends on the structure of the description. Since that structure is one of the goals of our research, any conclusions drawn from it are necessarily tentative. Another caveat is that the blocks world domain was deliberately chosen to minimize the complexity of the descriptive and expressive problems to be encountered.

As I mentioned previously, the hierarchical structure of the description is important in determining how much can be saved in replacing one frame with another. When a large, self-contained substructure such as a parallelogram face is essentially the same in the two descriptions, it is natural to preserve it as a unit rather than reconstructing it in the new frame. Even more than this is true in the transition to the three-face wedge: only a few parts of the top-level description need to be changed. The rest of the description remains the same.

In making the transition to the two-face wedge, the higher structures of the two descriptions are quite different, so less of the old description can be preserved. The interpretation of the observations remains the same, however: if two parts are considered connected by the parallelepiped frame, they are connected in the wedge frame, and the terms in which they are described are the same. Thus when the recognizer realizes it is looking at a wedge, it can remember what it saw when it thought the object was a parallelepiped. Even when the higher-level descriptive structure must be replaced, the recognizer need not look again at features it has already observed.

Here again we see an example where we are helped out by the good behavior of the domain, or at least of our view of the domain. Even when the frame changes, the interpretation of the observations remains much the same. This need not be true in domains with segmentation problems. For example, in speech recognition, changing the interpretation of one segment may affect the boundaries of the segment, requiring changes which ripple outward to neighboring hypotheses. A different set of techniques is required to state and evaluate hypotheses about domains where segmentation is an important problem.

V. SUMMARY

In this chapter we presented the idea of frames in a very intuitive way, outlining a number of desirable features of a representation for knowledge, and illustrating them with a specific example from the blocks world. A frame is a specialist in a small domain. It contains the knowledge necessary to create a description of an element of its domain from observed data. The features of such a description may be frames in their own right, representing a range of variation permitted in that domain. The frame for an object can have associated with it frames for actions which commonly affect that object, so that predictions can be made about required modifications to the description. The frame is capable of predicting unobserved features, and of using previous observations to refine its predictions. These predictions can guide the recognition process, and provide answers to questions before that process is complete. An observation which is inconsistent with the frame's expectations can suggest a better frame as a replacement. Much of the partially constructed description can be incorporated into the new frame, which continues the recognition process.

It is important to recognize the value of the intuitive model presented above. In a sense it is a "wish list" of desirable properties for a representation, but it is a list compiled with the problems of effective computability in mind. It will be many years before the technical problems implied by a frame theory can be precisely stated and solved. Such intuitions are therefore all the more important for providing a context in which current research can be viewed.

ACKNOWLEDGMENTS

I have received extremely valuable comments and suggestions from Marvin Minsky, Daniel Bobrow, and Carl Hewitt. Other people who have been very helpful include Michael Dunlavy, Mitch Marcus, Robert Metcalfe, Keith Nishihara, Don Norman, and Terry Winograd.

This research was done in part at the Artificial Intelligence Laboratory of the Massachusetts Institute of Technology. Support for the laboratory's artificial intelligence research is provided in part by the Advanced Research Projects Agency of the Department of Defense under Office of Naval Research contract N00014-70-A-0362-0003. I am also grateful to the Xerox Corporation for a stimulating summer spent at their Palo Alto Research Center.

REFERENCES

Dahl, O. J., & Hoare, C. A. R. Hierarchical program structures. In O. J. Dahl, E. W. Dijkstra, & C. A. R. Hoare (Eds.), *Structured programming*. New York: Academic Press, 1972.

Fahlman, S. E. A hypothesis-frame system for recognition problems (MIT-AI Working Paper 57). Cambridge: Massachusetts Institute of Technology, December, 1973.

Kay, A. *SMALLTALK, A communication medium for children of all ages*. Palo Alto, California: Xerox Palo Alto Research Center, Learning Research Group, 1974.

Minsky, M. A framework for representing knowledge. In P. Winston (Ed.), *The psychology of computer vision*. New York: McGraw-Hill, 1975.

Norman, D. A., & Bobrow, D. G. On data-limited and resource-limited processes. *Cognitive Psychology*, 1975, 7, 44-64.

Rubin, A. D. Hypothesis formation and evaluation in medical diagnosis (MIT-AI Technical Report 316). Cambridge: Massachusetts Institute of Technology, 1975.

Smith, B., & Hewitt, C. Towards a programming apprentice. *Artificial Intelligence and Simulation of Behaviour Summer Conference Proceedings*, University of Sussex, July, 1974.

Sussman, G. J. A scenario of planning and debugging in electronics circuit design (MIT-AI Working Paper 54). Cambridge: M.I.T., December, 1973.

Winograd, T. *Understanding Natural Language*. New York: Academic Press, 1972.

Winston, P. H. (Ed.). *The psychology of computer vision*. New York: McGraw-Hill, 1975.

FRAME REPRESENTATIONS AND THE DECLARATIVE/PROCEDURAL CONTROVERSY

Terry Winograd
Computer Science Department
Stanford University
Stanford, California

I. INTRODUCTION

Any discussion today of "the representation problem" is likely to entail a debate between proponents of *declarative* and *procedural* representations of knowledge. Sides are taken (often on the basis of affinity to particular research institutions) and examples are produced to show why each view is "right". Recently a number of people have proposed theories which purport to solve many representational problems through the use of something called "frames". Minsky (1975) is the most widely known example, but very similar ideas are found in Moore & Newell's (1973) MERLIN system, the networks of Norman & Rumelhart (1975), and in the schemata of D. Bobrow & Norman (Chapter 5). Parts of the notion are present in many current representations for natural language [see Winograd (1974) for a summary].

This chapter is composed of two distinct parts. In the first half I want to examine the essential features of the opposing viewpoints, and to provide some criteria for evaluating ideas for representation. The second half contains a very rough sketch of a particular version of a frame representation, and suggests the ways in which it can deal with the issues raised.

II. THE SIMPLE ISSUES

First let us look at the superficial lineup of the argument. It is an artificial intelligence incarnation of the old philosophical distinction between "knowing that" and "knowing how". The proceduralists assert that our knowledge is primarily a "knowing how". The human information processor is a stored program device, with its knowledge of the world *embedded* in the programs. What a person (or robot) knows about the English language, the game of chess, or the physical properties of his world is coextensive with his set of programs for operating with it. This view is most often associated with MIT, and is emphasized by Minsky and Papert (1972), Hewitt (1973), and Winograd (1972).

The declarativists, on the other hand, do not believe that knowledge of a subject is intimately bound with the procedures for its use. They see intelligence as resting on two bases: a quite general set of procedures for manipulating facts of all sorts, and a set of specific facts describing particular knowledge domains. In thinking, the general procedures are applied to the domain-specific data to make deductions. Often this process has been based on the model of axiomatic mathematics. The facts are *axioms* and the thought process involves *proof procedures* for drawing conclusions from them. One of the earliest and clearest advocates of this approach was McCarthy (see

McCarthy & Hayes, 1969), and it has been extensively explored at Stanford, and Edinburgh.

From a strictly formal view there is no distinction between the positions. Anyone who has programmed in languages like LISP has been forced into believing that "programs are data". We can think of the interpreter (or the hardware device, for that matter) as the only program in the system, and everything else as data on which it works. Everything, then, is declarative.

From the other end, we can view everything as a program. Hewitt, Bishop, & Steiger (1973) actually propose this view. A fact is a simple program which accepts inputs equivalent to questions like "Are you true?" and commands like "Assume you are true!". It returns outputs like "true" and "false", while having lasting effects which will determine the way it responds in the future (equivalent to setting internal variables.) Everything is a procedure. Clearly there is no sharp debate on whether a piece of knowledge *is* a program or a statement. We must go below these labels to see what we stand to gain in *looking at* it as one or the other. We must examine the mechanisms which have been developed for dealing with these representations, and the kind of advantages they offer for epistemology. In this entire discussion, we could divide the question into two aspects: "What kind of representation do people use?" and "What kind of representation is best for machine intelligence?" I will not make this distinction, first of all since the issues raised are quite similar, and second, since I believe that at our current stage of knowledge these questions are most profitably attacked as if they were the same.

A. The Benefits of Declaratives

Flexibility - Economy. The primary argument against procedural representation is that it requires that a piece of knowledge be specified by saying how it is used. Often there is more than a single possible use, and it seems unsatisfactory to believe that each use must be specified in advance. The obvious example is a simple universal fact like "All Chicago lawyers are clever." This could be used to answer a question like "Is Dan clever?" by checking to see if he is a Chicago lawyer. It might be used to decide that Richard is not from Chicago if we know he is a stupid lawyer, or that he is not a lawyer if he is a dim-witted Chicagoan. Each of these might be done in response to a question which is asked, or as a response to a new piece of information which has just been added. In a strictly procedural representation, the fact would have to be represented differently for each of these deductions. Each would demand a specific form, like "If you find out that someone is a lawyer, check to see if he is from Chicago, and if so assert that he is clever." Traditional logic, however, provides a simple declarative representation in the predicate calculus:

$$\forall(x) \; [\text{Chicagoan } (x) \; \& \; \text{Lawyer } (x) \Rightarrow \text{Clever } (x)]$$

The different uses for this fact result from its access by a general deductive mechanism. In adding this formula to the system, we do not have to anticipate how it will be used, and the program is therefore more flexible in the sorts of deduction it will be able to make. Associated with this flexibility is an obvious economy in making multiple uses from a single statement.

Understandability - Learnability. The simplicity of the declarative statement above is more important than just an economy of storage. It also has important implications for the ease of understanding and modifying the body of knowledge in a system. If the knowledge base is a set of independent facts, it can be changed by adding new ones, and the implications of each statement lie directly in its logical content. For programs, on the other hand, the implications lie largely in questions of how a routine is to be used, under what conditions, with what arguments, etc. Minor changes can have far-reaching effects on other parts of the program. In addition, it is not easy to split programs into independent subprograms. Thus a single "piece" of knowledge may be embedded as a line in a larger integrated program. Changing or adding is much more difficult. This issue of modification applies both to programmers trying to build a system, and to any sort of self-modification or learning.

Accessibility - communicability. Much of what we know is most easily statable as a set of declaratives. Natural language is primarily declarative, and the usual way to give information to another person is to break it into statements. This has implications both for adding knowledge to programs, and communicating their content to other people. Quite aside from how the eventual program runs, there may be important advantages in stating things declaratively, from the standpoint of building it and working with it.

B. The Benefits of Procedures

Procedural - Modeling. It is an obvious fact that many things we know are best seen as procedures, and it is difficult to describe them in a purely declarative way. If we want a robot to manipulate a simple world (such as a table top of toy blocks), we do it most naturally by describing its manipulations as programs. The knowledge about building stacks is in the form of a program to do it. Since we specify in detail just what part will be called when, we are free to build in assumptions about how different facts interrelate. For example, we know that calling a program to lift a block will not cause any changes in the relative positions of other blocks (making the assumption that we will only call the lift program for unencumbered blocks). In a declarative formalism, this fact must be stated in the form of a *frame axiom* which states something equivalent to "If you lift a block X, and block Y is on block Z before you start, and if X is not Y and X is not Z and X is unencumbered, then Y is on Z when you are done." This fact must be used each time we ask about Y and Z in order to check that the relation still holds. Note that this knowledge is taken care of "automatically" in the procedural representation because we have control over when particular knowledge will be used, and deal explicitly with the interactions between the different operations.

In using procedures we trade some degree of flexibility

for a tremendous gain in the ease of representing what we know about processes. This applies not only to obvious physical processes like moving blocks around, but equally well to deductive processes like playing games or proving geometry theorems.

Second Order Knowledge. One critical component of our knowledge is knowing about what we know and what we can know. We have explicit facts about how to use other facts in reasoning, and this must be expressed in our representation. These are not sophisticated logical tools, but straightforward heuristics. For example, if we want to generate a plan for getting to the airport, we know "If you do not see any obvious reason why the road should be impassable, assume you can drive." The catch is obviously the word "obvious". It is quite distinct from logical notions of truth and provability, referring to the complexity and difficulty of a particular reasoning process. Another such fact might be "The relation *NEAR* is transitive as long as you don't try to use it too many times in the same deduction."
It is theoretically possible to express second-order knowledge in a declarative form, but it is extremely difficult to do so outside the context of a particular reasoning process. In a procedural representation we can talk directly about things like the depth or duration of various computations, about the particular ways in which facts will be accesssed, etc.

The Need for Heuristic Knowledge. The strongest support for procedural representation comes from the fact that it works. Complex AI programs in all domains have a large amount of their knowledge built into their procedures. Those programs which attempt to keep domain-specific knowledge in a nonprocedural data base do so at the expense of limiting themselves to even more simplified worlds, and the simplest of goals. The obvious reason for this is that much of what we know about an area is *heuristic*. It is neither "simple facts" nor general knowledge about reasoning, but is of the form "If you are trying to deduce this particular sort of thing under this particular set of conditions, then you should try the following strategies." In theory, this knowledge could be kept separate and integrated with the declarative knowledge of a domain. In practice, it is not at all clear how a real system could do this. Most systems which have been based on a declarative formalism have only the general heuristics built into the interpreter, and do not make it easy to add domain-specific strategies. By putting knowledge in a primarily procedural form, we gain the ability to integrate the heuristic knowledge easily. The deduction process is primarily under control of the specific heuristic knowledge.

III. SOME UNDERLYING ISSUES

At this point it is tempting to look for a synthesis -- to say "You need both. Some things are better represented procedurally, others as declarative facts, and all we need to do is work on how these can be integrated." This reaction misses what I believe is the fundamental ground for the dispute. It is not simply a technical issue of formalisms, but is an expression of an underlying difference in attitude towards the problems of complexity. Declarativists and proceduralists differ in their approach to the duality between modularity and interaction, and their formalisms are a reflection of this viewpoint.

In his essay, The Architecture of Complexity, Simon (1969, p. 100) describes what he calls *nearly decomposable systems*, in which "...the short-run behavior of each of the component subsystems is approximately independent of the short-run behavior of the other components.... In the long run, the behavior of any one of the components depends in only an aggregate way on the behavior of the other components." One of the most powerful ideas of modern science is that many complex systems can be viewed as nearly decomposable systems, and that the components can be studied separately without constant attention to the interactions. If this were not true, the complexity of real-world systems would be far too great for meaningful understanding, and it is is possible (as Simon argues) that it would be too great for them to have resulted from a process of evolution.

In viewing systems this way, we must keep an eye on

both sides of the duality -- we must worry about finding the right decomposition, in order to reduce the apparent complexity, but we must also remember that "the interactions among subsystems are weak *but not negligible*". In representational terms, this forces us to have representations which facilitate the "weak interactions".

If we look at our debate between opposing epistemologies, we see two metaphors at opposite poles of the modularity/interaction spectrum. Modern symbolic mathematics makes strong use of modularity at both a global and a local level. Globally, one of the most powerful ideas of logic is the clear distinction between *axioms* and *rules of inference*. A mathematical object can be completely characterized by giving a set of axioms specific to it, without reference to procedures for using those axioms. Dually, a proof method can be described and understood completely in the absence of any specific set of axioms on which it is to operate. Locally, axioms represent the ultimate in decomposition of knowledge. Each axiom is taken as true, without regard to how it will interact with the others in the system. In fact, great care is taken to ensure the logical independence of the axioms. Thus a new axiom can be added with the guarantee that as long as it does not make the system inconsistent, anything which could be proved before is still valid. In some sense all changes are additive -- we can only "know different" by "knowing more".

Programming, on the other hand, is a metaphor in which interaction is primary. The programmer is in direct control of just what will be used when, and the internal functioning of any piece (subroutine) may have side effects which cause strong interactions with the functioning of other pieces. Globally there is no separation into "facts" and "process" -- they are interwoven in the sequence of operations. Locally, interactions are strong. It is often futile to try to understand the meaning of a particular subroutine without taking into account just when it will be called, in what environment, and how its results will be used. Knowledge in a program is not changed by adding new subroutines, but by a *debugging* process in which existing structures are modified, and the resulting changes in interaction must be explicitly accounted for.

If we look back to the advantages offered by the use of the two types of representation, we see that they are primarily advantages offered by different views toward modularity. The flexibility and economy of declarative knowledge come from the ability to decompose knowledge into "what" and "how". The learnability and understandability come from the strong independence of the individual axioms or facts. On the other hand, procedures give an immediate way of formulating the interactions between the static knowledge and the reasoning process, and allow a much richer and more powerful interaction between the "chunks" into which knowledge is divided. In trying to achieve a synthesis, we must ask not "how can we combine programs and facts?", but "How can our formalism take advantage of decomposability without sacrificing the possibilities for interaction?"

IV. STEPS TOWARD A MIDDLE

If the declarative and procedural formalisms represents endpoints on a spectrum of modularity/interaction, we should be able to see in each of them trends away from the extreme. Indeed, much current work in computing and AI can be seen in this light.

A. Modular Programming

One of the most prominent trends in computer programming today is toward some kind of *structured programming* as exemplified, say, by the work of Dahl, Dijkstra, & Hoare (1972). It represents a response to the tremendous complexities which arise when a programmer makes full use of his power to exploit interactions. Advocates maintain that understandability and modifiability of systems can only be maintained if they are forced to be "nearly decomposable". This is enforced by severely limiting the kinds of interaction which can be programmed, both in the flow of control and in the manipulation of data structures. This represents a move toward the kind of *local modularity* between the individual pieces of knowledge, and is carried to a logical extreme by Hewitt's actors.

Recent AI programming languages represent an attempt to achieve some *global modularity* within the programming context. [For a good overview, see Bobrow & Raphael (1974).] Through the use of pattern-directed call, and search strategies such as backup, they attempt to decouple the flow of control from the programmer. Procedures are specified whose meaning is in some sense free from the particular order in which they will be called, and the system has some sort of general mechanism for marshalling them in any particular case. So far, the general procedural knowledge is extremely primitive, and the potential modularity is rarely used. It is exploited much more fully in the *production systems* of Newell & Simon (1972). We can view production systems as a programming language in which all interaction is forced through a very narrow channel. Individual subroutines (productions) interact in only two ways, a static ordering and a limited communication area called the *short-term memory*. They can react to data left in the short-term memory and modify it as their means of communication. Their temporal interaction is completely determined by the data in this STM and a uniform ordering regime for deciding which productions will be activated in cases where more than one might apply. The orderings which have been most explored are a simple *linear* ordering of the productions, and a restricted system in which it is considered an error if there is not a unique production which matches the contents of short term memory at any time. Thus instead of the full ability to specify just what will happen when, the programmer can only determine the ordering, and the rest of the interaction is out of his hands. Of course, it is possible to use the STM to pass arbitrarily complex messages which embody any degree of interaction we want. The spirit of the venture, however, is very much opposed to this, and the formalism is interesting to the degree that complex processes can be described without resort to such tricks, maintaining the clear modularity between the pieces of knowledge and the global process which uses them.

B. Compiling Facts

Starting with a traditional logical-declarative system, Sandewall (1973) attempts to build in some of the specific control interactions which make programs effective. He compiles declarative information into *operators* in which specific decisions are embedded about what knowledge should be called into play when. These are then included as part of a system which in many ways is similar to GPS, the intellectual forerunner of the production systems. Another approach from this direction is Sussman's (1973) attempt to combine the effectiveness of procedures with the ease of modification which comes from modularity. His system contains both declarative and procedural knowledge, but combines them by being an *active programmer*. There is a body of declarative data about the specific subject domain, but it is not used directly in this form. As each piece is added, whether as a statement from a teacher, or from an experience in the model-world, it is perused by a programmer-debugger. General knowledge about procedures is brought into play to decide just how the new knowledge should be integrated into the domain-specific programs, and how the resulting interactions might be anticipated and tested. Thus what the system knows may be decomposed into "procedure" and "domain fact" modules, but these are internally combined into a procedural representation.

V. A FIRST ATTEMPT AT SYNTHESIS

Recently much excitement has been generated by the idea of a representational format called "frames" which could integrate many of the new directions described in the previous section. So far, this work is in a beginning state of development, and none of the available papers work out the actual implementation of such a scheme and its application to a significant set of problems. Therefore, most of what can be said is at the level of general system criteria, and ideas for organization.

In this section I give a simple example of a system which represents knowledge in a frame-like notation. I warn the reader strongly that this does not purport to be an explanation of what frames "really are", or to represent anyone else's understanding of what they should be. Many different issues are still unsettled (many more perhaps unrecognized), and it will be a long time before any agreements on notation and operation can be reached. The notation also does not represent a worked-out design, but is intended to be suggestive of the necessary formalism. This example is specifically chosen to *avoid* many of the most interesting problems, in order to gain at least a small foothold for attacking them sensibly. Thus it oversimplifies and misrepresents my own views of frames as well. This version grew up in the same environment as Minsky's (1975) frames, but with a slightly different emphasis since they were initially applied with a view toward natural language rather than vision. Their development over the last year has been strongly influenced by discussions with Andee Rubin at MIT, and Daniel Bobrow at Xerox PARC. Further work is continuing in conjunction with Bobrow, leading toward the specification of the frame formalism and the development of reasoning programs which use it.

With this caveat clearly in mind, let us look at the problem of understanding the connection between days, dates, and numbers. The context for this problem can be best understood by imagining some sort of program (or person) acting as an office assistant in matters such as scheduling. The assistant must be able to do things like writing schedules, accepting information about events to be scheduled, and accepting facts about dates in a natural input. This does not necessarily mean natural language, but in a form whose structure corresponds roughly to that which a person might use to another person.

A. The Generalization Hierarchy

We first note that the system would have a set of internal concepts appropriate to the subject matter. These might include things such as those in Fig. 1. These concepts are arranged in a *generalization hierarchy*, a structure of *isa* links connecting concepts to those of which they are *specializations*. This hierarchy contains both specific objects (like *July 4, 1974*) and general objects (like *day, holiday,* and *Thursday*). I will avoid discussing the problems inherent in the combination of these in a single hierarchy, as that deserves at least another entire paper.

Associated with each node in this hierarchy is a *frame* tying together the knowledge we have of that concept. Often there will also be a particular English word or phrase associated with it, but that is not a necessary condition. Reference within the system can be made by using the internal concept names as shown in the hierarchy.

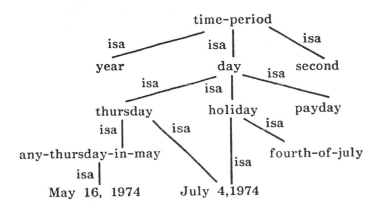

Fig. 1. A generalization hierarchy.

It is clear that this cannot be a simple tree, since often two different generalizations apply to the same specific concept. *July 4, 1974* is both a *fourth-of-july* and a *Thursday*. The use of this hierarchy is primarily through *inheritance of properties* Any property true of a concept in the hierarchy is implicitly true of anything linked below it, unless explicitly contradicted at the lower level. This is an old idea, to be seen, for example, in the work of Raphael and Quillian (in Minsky, 1968). At times it has been used to cover too much; however, as one of many deductive mechanisms, it is particularly efficient and intuitively reasonable. Again, many issues can be raised about the problem of "multiple isa links". If two generalizations apply, how do we choose the appropriate one to look for the properties we want, and how do we resolve

disagreements between them?   For the purposes of this
elementary discussion, we will ignore these (which of course
does not make them go away for good).

B. Description and Classification

In the previous section we used the word
"generalization" to describe the relationship between frames
linked in the hierarchy. Another way to think of it would
be as a system of *classification*. Each frame in fact
represents a class of objects, and an *isa* link connects a
class to some superclass which properly contains it. From
an operational viewpoint it seems more useful to think of
this as a hierarchy of *descriptions*. Of course there is a
duality between the idea of descriptions and the idea of the
classes of objects to which they apply. In making use of
these frames, we will be specifying and modifying
descriptions, not dealing with the sets of objects (mental or
physical). Thus we will use the notion that a general
description can be *further specified* to any one of a number
of more specific ones. At the bottom we have descriptions
which apply to unique objects in the system's model of the
world. In attaching particular properties or facts to a node
of this hierarchy, we are saying "Anything to which the
description at this node applies also has these additional
properties."
This operation of applying classifications is one of the
basic modes of reasoning. We decide on the basis of
partial evidence that some particular object with which we
are concerned belongs to a known class (i.e., there is a
particular frame for it in our hierarchy). On the basis of
that decision, we can apply a whole set of additional
knowledge associated with that frame. We can use the
frame to guide our search for the specific facts associated
with the object, or to make assumptions about things that
must be true of it without checking specifically. [Minsky
(1975) has an extended discussion of this vital aspect of
frames as does Kuipers, Chapter 6.].
This type of reasoning can be extended to handle things
more like analogy. In attempting to apply a full
description to an object which would not normally be in

the corresponding class, the frame focuses our attention on those particular properties or facts which are applicable, and those which are contradictory.

C. Important Elements

What then are the elements attached to a description? In a simple predicate scheme, our classifications correspond to the set of predicates. All axioms (or clauses) containing a predicate are attached implicitly by the fact that they contain the symbol. Any statement which contains the predicate $day(x)$ says something about the concept *day*. In a semantic net system, the attachments are explicit, and essentially constitute all the links from the concept to anything else. The frame idea adds an important additional element to this -- the idea of *importance* or *centrality*. For each frame, there is a set of other frames marked as *important elements* having a specific relation to it. These are what Minsky (1975) calls "slots", Newell calls "components of the beta-structure", etc. The details of how these are determined and treated is probably the key area of frame theory, and the one on which there is the least agreement. Again with the warning that this oversimplification does not represent anyone's full views (including mine), I will pare away most of the interesting problems and go on a set of intuitive but unexplored assumptions for this particular case.

Associated with the frame for *day*, the important elements (or *IMPS*) might be, for example, *year, month, day-numbers, day-of-week, sequence-number* and *ASCII-form*. Immediately we note that the idea of important elements must be sensitive to what we are doing. The idea of *day* would have very different elements if we were thinking of it as a typical sequence of events for a person (getting up, going to work,....), as an astronomical phenomenon (period of rotation, orientation of axis,...), as a set of natural events (sunrise,...,sunset, night), etc. We need separate frames for things like *day viewed as a calendar object*, *day viewed as an event sequence*, etc. Clearly more mechanism is needed to explain the connections between these (how do they relate in more ways than the fact they use the same word), and the ability to choose the appropriate one for a given context. The idea of attaching IMPs to frames is not intended to avoid the more usual types of deduction and association, but to augment it by providing some kind of formalism for talking about the importance of certain ideas and relations with regard to a central concept (frame) within a context.

TABLE I.
Descriptions of the IMPs for "day"

IMP	DESCRIPTION
Year	Integer
Month	Month-name
Day-number	(Integer range (interval min 1 max 31))
Day-of-week	Weekday-name
Sequence-number	Integer
ASCII-form	(Integer length 6 structure (concatenated-repetition element (integer length 2) number 3)))

When we further specify a frame (that is, move down the generalization hierarchy) we also further specify the IMPs that go with it. This makes sense only if we think of each IMP as being another frame (which in turn fits into the hierarchy, and so on recursively). Thus the IMPs associated with *day* might be initially filled by the descriptions of Table I. These of course imply a much more extended set of frames to cover things like "integer", "string", "6", the relation "length", etc. This knowledge will be common to many of the knowledge domains represented in a computer system, and one of the basic prejudices of this approach is that in looking at the knowledge of any part of the system we must do it in terms of its dependency on other parts. Our world knowledge is all tied together, and in any particular area we make use of our more general concepts. We have adopted a simple notation in which a frame is given by putting the name of its generalization (the thing to which it has an *isa* link) followed by a list of IMP names and their contents. If the appropriate frame is a simple named one, whose IMPS are not further specified, that name is used directly. This should definitely not be interpreted in the usual LISP sense of atoms and functions, and is also not isomorphic to a kind of straightforward atom-property list notation. Since the problems of all this do not arise at the level of detail used in the discussion below, they will be ignored for now [see Moore & Newell (1973) for a discussion of some of them].

At first glance these descriptions look like traditional computer language data types--integer, string, etc. Indeed, data types are one simple example of a shallow generalization hierarchy, but do not extend nearly far enough. We must be able to provide descriptions at any level of detail rather than choosing from a small finite set of categories. Newer computer language formalisms are beginning to move in this direction. [See Dahl, Dijkstra, & Hoare, 1972; Cheatham & Wegbreit, 1973.] Also, it should be pointed out that the choice of descriptions is not strictly determined. The fact that the standard ASCII representation of a date is a six-digit integer is simple. But the fact that this integer is a sequence of three two-digit integers is less straightforward. Clearly any six-digit integer can be viewed in this way, but the fact that this description is used explicitly is a statement that this way of looking at it is in fact relevant and important for the purposes of this frame. One of the key points of the notation is that it allows (encourages) us to include facts which although in themselves "epistemological" (to use McCarthy's term) give us heuristic clues by their very presence in the representation. We want to represent not just the bare essence of what we know, but also the scheme of how we think about it. Here we are treading on that middle ground between declarative and procedural knowledge, by selecting our declarative statements with specific concern for how they will be used.

D. Relations between IMPs

We can now express a frame for a particular day as in Table II or a more specific but still general day as in Table III.

TABLE II.
Further Specification of IMPs for "July 4, 1974"

IMP	DESCRIPTION
Year	1974
Month	July
Day-number	4
Day-of-week	
Sequence-number	
ASCII-form	740704

TABLE III.
Further Specification IMPs for "Any Thursday in May"

IMP	DESCRIPTION
Year	
Month	May
Day-number	
Day-of-week	Thursday
Sequence-number	
ASCII-form	

In each case, we have filled in those IMPs which correspond to the "natural" way of describing that day. The *filler* for each IMP is a further specification of the same IMP in the *day* frame. For example, *4* is a further specification of *(integer range (interval min 1 max 31))* which in turn is a further specification of the general frame for *integer* in the part of the generalization hierarchy dealing with numbers. It is obvious that a more extended formalism is needed to express what we know about dates. The set of IMPs is not independent. Quite the opposite -- it is the interrelation between these important elements that provides most of the useful knowledge connected with the frame. Thus we need a notation which allows specific reference to IMPs associated with a frame. Newell seems to want to avoid this in his beta-structures, but it is not clear that this can be done at all, and certainly not without paying an untenable price in combinatorial explosion. We have chosen to represent an element of a frame as a path of IMP names implicitly beginning with the frame in which the path appears. Once again there are many issues buried here about variable binding, scoping, use of pointers versus names, etc., and the notation is far from settled. A list beginning with a "*!*" is used to indicate such a reference path. Thus in Table IV we include additional knowledge about our general concept of *day*, such as the fact that the year is actually the digits "19" concatenated with the first two digits of the ASCII form, or that we can expect each month name to have associated with it an IMP named *length*, and that the range of the *day-number* is determined by it. In adding this information, we have used material from still more of the fundamental set of frames used by a problem solver, such as *1-1 correspondence* and *position-in-list*.

E. Procedural Attachment

So far we have been building a declarative data structure. It contains somewhat more information than a simple set of facts, as it has grouped them in terms of importance, and made some of them implicit in the hierarchy. We can deduce that "The ASCII code for July 4

TABLE IV.
Relations between IMPs

IMP	DESCRIPTION

Year	(Integer
	  structure (concatenation
	             first "19"
	             second
	             (! ASCII structure first)))

Month	(Month-name
	  (position-in-list
	    list "January, February,...,December"
	    element (! month)
	    number (! ASCII structure second)))

Day-number	(Integer
	  range (interval
	       min 1
	       max (! month length)))

Day-of-week	(Weekday-name
	  (position-in-list
	    list "Sunday, Monday,...,Saturday"
	    element (! day-of-week)
	    number (integer
	       range (interval min 1 max 7)))
	  (1-1 correspondence
	    set1
	      (! day-of-week position-in-list number)
	    set2 (quotient-mod-7
	         dividend (! sequence-number))))

Sequence-number	Integer

ASCII-form	(Integer
	  length 6
	  structure (concatenated repetition
	        element (integer length 2)
	        number 3)))

is a six-digit integer." using only the hierarchical link, while deciding "Its final four digits are 0704." would involve more complex computations, using the pointers and IMPs.

At this point we might try to describe a general deductive mechanism which made use of this elaborated structure, taking advantage of the additional information to help guide its computations. It seems possible to do so. The special use of the generalization hierarchy does not really provide any different power than that which could be expressed by a series of simple axioms like

$\forall(x)$ Holiday(x) $\Rightarrow$ Day(x).

There have been various schemes to use groupings of axioms to guide the theorem prover in selecting which to use (although it might be very difficult to prove things formally about its behavior). To the degree such a scheme succeeded, we would have been successful at embedding heuristic knowledge into the declarative structure. I believe, however, that to be useful, a representation must make much more specific contact with procedures.

In one sense, the idea of frame structure provides a framework on which to hang procedures for carrying out specific computations. Table V indicates some of the procedures which might be attached to our frame for *day*. The notations *TO-FILL* and *WHEN-FILLED* correspond loosely to the antcedent/consequent distinction of Planner-like formalisms, and indicate the two fundamental reasons for *triggering* a computation. The mechanism for doing this should, however be more general than the syntactic pattern match which controls the calling of procedures in these languages. Triggering should be based on a more general notion of mapping [see Moore & Newell (1973) and Minsky (1975) for different versions of this.] We will describe later one aspect of how this interacts with the abstraction hierarchy.

The procedures attached to a frame or IMP are not necessarily equivalent to the factual information in the same place. One obvious example is the procedure *use-calendar* associated with *day-of-week*. Our system would have a special algorithm which can take as inputs the day-number, month, and year, and produce as output a week-day. Its steps might be paraphrased as "Find a calendar. Find the page corresponding to *month-name*. Find the number corresponding to *day-number* on the page. Look at

TABLE V.
Possible Procedures to Attach to the "Day" Frame

IMP	PROCEDURE
Year	
Month	
Day-number	WHEN-FILLED
	(CHECK-RELATION day-number, month)
Day-of-week	TO-FILL
	(APPLY calendar-lookup
	TO year, month, day-number)
	(APPLY anchor-date-method
	TO year, month, day-number)
Sequence-number	
ASCII-form	WHEN-FILLED
	(FILL year, month, day-number)

the top of that column." This is the algorithm most of us use most of the time for this computation. It depends on a batch of additional knowledge about calendars and how they are printed, pages, columns, etc., but in using this procedure we do not worry about why calendars work -- only the direct steps needed. At a deep level it is based on the facts listed in Table IV, but that is far removed from the procedure itself. This is an important element of our frame notation -- there is a large degree of redundancy between the procedural and declarative knowledge. Many procedures are specific ways to deduce things which are implicit in the facts. Many facts are essentially statements about what the corresponding procedures do. Neither one is the "fundamental" representation--in any individual case one or the other may be learned first, one or the other may be used in more circumstances.

There may be a number of different procedures attached to any frame or IMP. In this example, we have another procedure for calculating dates, based on the use of an "anchor date". If asked "What day is June 28 this year?" I may perform a computation whose trace would be: "I know that May 12 is a Sunday, so the 19th is a Sunday, the 26th is a Sunday, the 33rd is a Sunday, but May has only 31 days, so June 2 is a Sunday, so the 23rd is, so the 28th is a Friday." This procedure involves a specific algorithm for beginning with some known date, stepping forward by 7 (or multiples of 7), accounting for the number of days in a month, and finally counting backward or forward from a known day of the week. (This last calculation often involves the use of peripheral digital devices (fingers) as well.) This procedure is much more closely tied to facts like the connection of the weekdays to the sequence of day numbers, but is not simply an application of them. The organization of these facts into a specific algorithm represents the addition of relevant knowledge, and its inclusion is not purely redundant. In operating it will also make fairly direct use of facts which are in the simple declarative form, such as that the *length* of May is 31. This new procedure is more generally applicable than the calendar one -- it works when we don't have a calendar at hand, and even works if we say "Imagine that May had 32 days this year. Then what day would June 28 be on?"

We can view the procedures as covering a whole spectrum, from very specific ones (attached to frames near the bottom of the hierarchy) to highly general ones nearer the top. If we did not have a specific procedure for finding the name of the month from the ASCII representation, we could look at the facts and try applying whatever procedures were connected to the frame for *1-1 correspondence*. In fact we should have a number of different frames which are further specifications of *1-1 correspondence* for different cases. There may be only an abstract mathematical correspondence in which we know only of its existence, but not how to compute the actual correspondence between particular elements; a functional correspondence in which given a domain element we have a direct way of computing a range element, but not vice versa; a testable-pair correspondence in which we can test whether two elements correspond, but have no good way to search for one, etc. This enrichment of the set of terms used in the facts is crucial to integrating them properly with the procedures. The set of notations which

corresponds loosely to the quantifiers and connectives of
logic would be much larger and richer, involving a
correspondingly expanded set of inference rules. Again,
this does not increase the abstract logical power, but allows
us to include much more procedurally related information
("What can you do with this fact") into the declarative
notation.

The procedural information attached to frames may be
very specific, or may be simply a guide of what to do, like
the statment (FILL month-name, year, day-number) in
Table V. This does not indicate the specific procedures to
be used for each of these, but is rather a piece of *control-
structure* information. It tells the system that on filling
the ASCII representation, it might be a good strategy to do
these other computations. The system then must look for
procedures attached to those paricular IMPs (or
generalizations of them) to do the work. One important
area to be explored is the way in which this sort of
explicit control structure can be included without paying a
high price in loss of modularity.

F. Problems of Learning and Modularity

One important aspect of procedural attachment is that it
should be dynamic -- the basic ways the system learns are
by adding new facts (either from being told or by inducing
a generalization), and by creating new procedures based on
applying general procedures to specific facts. This is the
type of learning described by Sussman (1973). We can
think of his work as beginning with a very general set of
procedures (near the top of our hierarchy) and a set of
specific facts. The outcome is a set of more specific
procedures associated with the particular tasks to be done
at all levels of the hierarchy. The frame framework does
not attack the details of this kind of debugging, but
provides a way of representing what is going on and
integrating the use of the results. Similarly we can think
of the task of *automatic programming* as "pushing procedures
down the hierarchy". Given very general procedural
knowledge and declarative facts about further specified
cases, the task is to write the procedures suited to those
cases.

Returning to the issue of modularity, we find that it has been attacked directly by adding a whole new layer of structure. Instead of taking a modular view (each fact is independent) or a highly integrated view (everything knows enough about the others to know how to use them) we have created a set of mechanisms for allowing the writer (whether human or program) to impose a kind of modularity through the decision of what will be included in a frame, and how the procedures will be attached. Modularity then becomes not a fixed decision of system design, but a factor to be manipulated along with all the other issues of representation.

Most of what the system knows is included in *both a modular and an integrated form.* The procedures for learning and debugging continually use general knowledge of programming to take the individual facts and combine them into the specific integrated procedures which do most of the system's deductions. Faced with a problem for which no specific methods are available, or the ones available do not seem to work, the system uses the specific facts with more general methods. There is no sharp division between specific and general methods, since there is an entire hierarchy of methods attached at all levels of the generalization hierarchy for the concepts in the problem domain. The most critical problem for the representation is to make it possible for this shifting between levels of knowledge to occur smoothly, without demanding that the programmer anticipate the particular interactions.

VI. CONCLUSION

This formalism has clearly achieved one goal it set out to do: to blur many of the distinctions such as declarative/procedural, and heuristic/epistemological in the discussion of representations. It is yet to be seen whether this blurring will in the end clarify our vision, or whether it will only lead to badly directed groping. Many more issues are raised by it than are immediately solved, and for each solution there are many more problems. Hopefully this book will be one step on the way to sorting out what is useful and providing a new generation of representational tools for artificial intelligence.

REFERENCES

Bobrow, D. G., & Raphael, B. New programming languages for artificial intelligence research. *Computing Surveys*, 1974, *6(3)*, 153-174.

Dahl, O. J., Dijkstra, E., & Hoare, C. A. R. *Structured programming*. New York: Academic Press, 1972.

Hewitt, C., Bishop, P., & Steiger, R. A universal modular ACTOR formalism for artificial intelligence. *Proceedings of the Third International Joint Conference on Artificial Intelligence*, 1973, 235-245.

McCarthy, J., & Hayes, P. Some philosophical problems from the standpoint of artificial Intelligence. In B. Meltzer and D. Michie (Eds.), *Machine Intelligence 4*. Edinburgh, 1969, 463-502.

Marvin Minsky (Ed.), *Semantic Information Processing* Cambridge: MIT Press, 1966.

Minsky, M. A framework for representing knowledge. In Winston, P. (Ed.), *The psychology of computer vision*. New York: McGraw-Hill, 1975.

Minsky, M., & Papert. S. Progress Report. Massachusetts Institute of Technology, Artificial Intelligence Laboratory, 1972.

Moore, J, & Newell, A. How can MERLIN understand?. In Gregg (Ed.), *Knowledge and cognition*. Baltimore, Md.: Lawrence Erlbaum Associates, 1973.

Newell, A., & Simon, H. A. *Human Problem Solving*. Prentice Hall, 1972.

Sandewall, E. Conversion of predicate-calculus axioms, viewed as non-deterministic programs, to corresponding deterministic programs. *Proceedings of the Third International Conference on Artificial Intelligence*, 1973, 230-234.

Simon, H. The architecture of complexity. In *The Sciences of the Artificial*. MIT Press, 1969.

Sussman, G., *A computational model of skill acquisition* (MIT-AI TR 297), 1973.

Winograd, T. *Understanding Natural Language*, New York: Academic Press, 1972.

Winograd, T., *Five lectures on artificial intelligence* (Stanford AI-Memo-246). September 1974.

NOTES ON A SCHEMA FOR STORIES

David E. Rumelhart

University of California, San Diego
La Jolla, California

I. INTRODUCTION

Just as simple sentences can be said to have an internal structure, so too can stories be said to have an internal structure. This is so in spite of the fact that no one has ever been able to specify a general structure for stories that will distinguish the strings of sentences which form stories from strings which do not. Nevertheless, the notion of "well-formedness" is nearly as reasonable for stories as it is for sentences. Consider the following examples:

(1) Margie was holding tightly to the string of her beautiful new balloon. Suddenly, a gust of wind caught it. The wind carried it into a tree. The balloon hit a branch and burst. Margie cried and cried.

211

(2) Margie cried and cried. The balloon hit a branch and burst. The wind carried it into a tree. Suddenly a gust of wind caught it. Margie was holding tightly to the string of her beautiful new balloon.

Here we find two strings of sentences. One, however, also seems to form a sensible whole, whereas the other seems to be analyzable into little more than a string of sentences. These examples should make clear that some higher level of organization takes place in stories that does not take place in strings of sentences. The purpose of this chapter is to illustrate that point, to develop some notions of the sorts of structures that might be involved, and to illustrate how these structures can be used to produce cogent summaries of stories.

To begin, it is clear that simple sentences are not the highest level of structured linguistic input. Sentences themselves can serve as arguments for higher predicates and thus form more complex sentences. For example,

(3) Margie knew that her balloon had burst.

Here we have one sentence about the bursting of Margie's balloon embedded as the argument of a higher verb. Sentences such as these, of course, occur with high frequency. Another case in which sentences occur as arguments of higher predicates is

(4) Margie cried and cried because her balloon broke.

In this case the predicate "because" takes two sentences as arguments. Now consider the following pair of sentences:

(5a) Margie's balloon broke.
(5b) Margie cried and cried.

It seems clear that the sentence pair (5a) and (5b) have almost the same meaning as (4) and ought therefore to have the same underlying structure. Thus if we are to understand correctly (5a) and (5b) we must infer the causal relationship between the propositions. This, I suspect, is but a scratch on the surface of the kinds of "suprasentential" relationships that are implied and understood in ordinary discourse. In particular, I suggest that the structure of stories is ordinarily more than pairwise relationships among sentences. Rather, strings of sentences combine into psychological wholes. In the following section I explore the nature of these wholes and propose a simple story grammar which accounts for many of the salient facts about the structure of simple stories and which will serve as the basis for a theory of summarization.

II. A SIMPLE STORY GRAMMAR

A. The Grammar Rules

In this section I will develop a grammar which I suggest accounts in a reasonable way for the structure of a wide range of simple stories. The grammar consists of a set of syntactical rules which generate the constituent structure of stories and a corresponding set of semantic interpretation rules which determine the semantic representation of the story. The symbol "+" is used to form two items in a sequence; the symbol "|" is used to separate mutually exclusive alternatives. A "*" following a structure name indicates one or more of those units; for example, A* is one or more As.

Rule 1: Story -> Setting + Episode

The first rule of our grammar says simply that stories consist of a Setting followed by an Episode. The Setting is a statement of the time and place of a story as well as an introduction to its main characters. The Setting corresponds to the initial section of stories such as:

Once upon a time, in a far away land, there lived a good king, his beautiful queen, and their daughter Princess Cordelia . . .

The setting is usually just a series of stative propositions, often terminated by phrases such as:

One day, as Princess Cordelia was walking near the palace . . .

In the story illustrated in the first example--the Margie story--the setting consisted of the sentence:

Margie was holding the string of her beautiful new balloon.

The remainder of the story is an Episode. The simple semantic rule corresponding to Rule 1 is:

Rule 1': ALLOW (Setting, Episode)

Semantically, the setting forms a structure into which the remainder of the story can be linked. It plays no integral part in the body of the story and under certain conditions can be eliminated without adversely effecting the story. In such cases, the characters and their relevant characteristics must be introduced in the body of the story.

Rule 2: Setting -> (State)*

Rule 2 simply expresses the assumption that settings consist of a set of stative propositions.

Rule 2': AND (State, State, . . .)

Semantically, the states are represented as a set of conjoined propositions entered into the data base.
The first real substantive rule in our rewrite for Episode is:

Rule 3: Episode -> Event + Reaction

Episodes are special kinds of events which involve the reactions of animate (or anthropomorphized) objects to events in the world. The episode consists merely of the occurrence of some event followed by the reaction of the hero of the episode to the event. Our semantic rule corresponding to Rule 3 is:

Rule 3': INITIATE (Event, Reaction)

That is, the relationship between the external event and the hero's reaction is one that I call INITIATE. I have taken the term from Schank (1973), although my use is slightly different from his. I use the term INITIATE to represent a kind of causal relationship between an external event and the willful reaction of a thinking being to that event. In the Margie story illustrated in (1), I assume that the relationship between Margie's crying and the breaking of her balloon is the INITIATE relationship. Presumably, the crying is mediated by an internal mental response such as "sadness".

Event is the most general category of our entire grammar. The following rule expresses the structure of an event:

Rule 4:   Event -> {Episode | Change-of-state | Action | Event + Event}

Thus an Event can be any of the alternatives, an episode, a simple change of state, or an action that people carry out. All are special kinds of events. Furthermore, a sequence of events also can constitute an event. The first three parts of Rule 4 require no semantic interpretation rules. Our semantic rule corresponding to the fourth rewrite for event is:

Rule 4':   CAUSE (Event, Event) or ALLOW (Event, Event)

The rule states that a sequence of two events can either be interpreted as one event CAUSE a second event or they can be interpreted as the first event ALLOW the second. The term CAUSE is used when the relationship between the events is one of physical causation as in the balloon hitting the branch causing the balloon to break in the Margie story. (The CAUSE predicate is similar to Schank's (1973) RESULT.) ALLOW is a relationship between two events in which the first makes the second possible, but does not cause it; thus the relationship between the wind catching the balloon and the wind carrying it into the tree I would say is ALLOW. (Here again my usage of ALLOW is clearly closely related to Schank's ENABLE, but is probably not identical.)

Rule 5:   Reaction -> Internal Response + Overt Response

Thus a reaction consists of two parts, an internal and an overt response.   The semantic relation between these two responses is:

Rule 5':   MOTIVATE (Internal Response, Overt Response)

MOTIVATE is the term used to relate thoughts to their corresponding overt actions.

Presumably there are a large variety of types of internal responses.   The two most common, however, seem to be emotions and desires.   Thus we have:

Rule 6:   Internal Response -> {Emotion | Desire}

Presumably other internal responses can be aroused, but in the stories I have analyzed these two have been sufficient. The overt response is, of course, semantically constrained to be a plausible response for our particular internal response.

B. An Example

Before continuing the explication of the grammatical rules, it may be useful to illustrate the structure--semantic and syntactic--generated by these rules for the Margie story.   Fig. 1 illustrates the analysis.   Fig. 1a gives the units of the Margie story.   Fig. 1b illustrates the tree generated by application of the syntactic rules.   Fig. 1c illustrates our semantic structures.

(1) Margie was holding tightly to the string of her beautiful new balloon.
(2) Suddenly, a gust of wind caught it
(3) and carried it into a tree
(4) It hit a branch
(5) and burst.
(6) [sadness]
(7) Margie cried and cried.

Fig. 1a. The units of the Margie story.

In the semantic structures the convention is followed
that predicate names written in the ovals and the
arguments of the predicates are pointed to by arrows. The
propositions which are the units of the story are simply
numbered. Note also that inferred propositions are written
in the body of the story in square brackets. This simply
facilitates our reference to these propositions. In the
semantic structures, I follow the convention that the left-
hand arrow points to the "causal event" and the right-hand
arrow points to the "caused event". This notation is
similar to that given in Rumelhart, Lindsay, & Norman
(1972) and in Norman, Rumelhart, & the LNR research
group (1975).

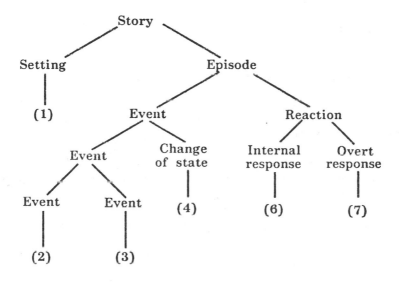

Fig. 1b.   The syntactic structure of the story.

I will return, in a later section of this chapter, to a
discussion of the role of our semantic and syntactic
structures in story processing. At this point, however, we
return to the development of our story grammar.

David E. Rumelhart

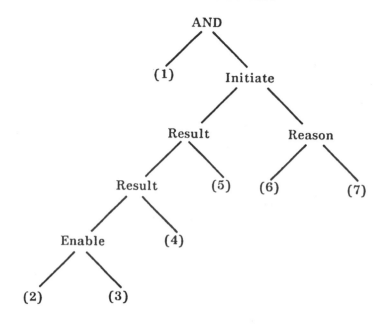

Fig. 1c. The semantic structure of the story.

In the case illustrated above, the overt response node had a trivial expansion--namely it was interpreted simply as an action. Other times it has a more complex expansion. In particular, when the internal response is a desire, the overt response is an attempt or rather a series of attempts at obtaining that desire. Thus we have the following syntactic rules:

Rule 7: Overt Response -> {Action | (Attempt)*}

Corresponding to Rule 7 we have the following semantic interpretation rule:

Rule 7': THEN (Attempt, Attempt, . . .)

The THEN relation is a predicate of any number of arguments of which the nth argument occurs prior to the n-1st argument in time.

Now, attempts too have their own internal structure:

Rule 8: Attempt -> Plan + Application

TABLE I.
Syntactic Rules and Semantic Interpretation Rules

---

(1) Story -> Setting + Episode
    => ALLOW (Setting, Episode)
(2) Setting -> (States)*
    => AND (State, state,......)
(3) Episode -> Event + Reaction
    => INITIATE (Event, Reaction)
(4) Event -> {Episode | Change-of-state | Action | Event + Event}
    => CAUSE (Event$_1$, Event$_2$) or ALLOW (Event$_1$, Event$_2$)
(5) Reaction -> Internal Response + Overt Response
    => MOTIVATE (Interval-response, Overt Response)
(6) Internal Response -> {Emotion | Desire}
(7) Overt Response -> {Action | (Attempt)*}
    => THEN (Attempt$_1$, Attempt$_2$,......)
(8) Attempt -> Plan + Application
    => MOTIVATE (Plan, Application)
(9) Application -> (Preaction)* + Action + Consequence
    => ALLOW (AND(Preaction, Preaction,..),
       {CAUSE | INITIATE | ALLOW} (Action, Consequence))
(10) Preaction -> Subgoal + (Attempt)*
    => MOTIVATE [Subgoal, THEN (Attempt,......)]
(11) Consequence -> {Reaction | Event}

In other words, an attempt consists of two parts, the development of a plan (to obtain the desire) and the application of that plan, Rule 8' is the semantic interpretation for Rule 8:

Rule 8': MOTIVATE (Plan, Application)

In words, Rule 8' states that the Plan MOTIVATES the application of the plan. The application consists of three parts; an optional series of preactions optionally followed by an action, and a consequence of that action. Thus we have:

Rule 9: Application -> (Preaction)* + (Action + consequence)

TABLE II.
Definitions of Semantic Relationships

AND | A simple conjunctive predicate of any number of arguments.

ALLOW | The relationship between an event which made possible, but which did not directly cause a second event. Thus for example, going to the store ALLOWS but does not cause me to buy some bread.

INITIATE | The relationship between an external event and the willful reaction of an anthropomorphized being to that event. Thus for example, an angry lion escaping from a cage in front of me INITIATES my being afraid and running for safety.

MOTIVATE | The relationship between an internal response and the actions resulting from that internal response. Thus for example, my being afraid MOTIVATES my running to safety.

CAUSE | The relationship between two events in which the first is the physical cause of the second. Thus for example, the baseball striking the bat CAUSES the baseball flying towards the outfield.

THEN | The relationship which holds among a temporally sequenced set of events. The first argument occurs prior in time to the second, the second prior to the third, etc.

The semantic interpretation of this rule is:

Rule 9': ALLOW [AND (Preaction, . . .) {INITIATE | CAUSE | ALLOW} (Action,Consequence)]

In words, Rule 9' states that the temporally sequenced set of preactions ALLOW the action which either INITIATES, CAUSES or ALLOWS the Consequence. Which of these three semantic relationships are appropriate depends on the nature of the action and its consequence. If none of the preactions are present, the semantics of the embedded proposition become the interpretation for the entire rule. Furthermore, in some stories one or more of the preactions fail and thus the action itself may never appear. In such cases the consequence also fails to appear and the semantics of Rule 9' is simply the temporal sequence of those preactions which do occur.

Two rules remain to complete the development of the story grammar. These are the rules for rewriting preactions and for rewriting consequences:

Rule 10:   Preaction -> Subgoal + (Attempt)*

Rule 10': MOTIVATE [Subgoal, THEN (Attempt, . . .)]

Rules 10 and 10' state that preactions consist of the development of a subgoal which serves to MOTIVATE a series of attempts to obtain that subgoal. The consequence is a very general category. Consequences, it seems, can be any sort of event resulting from the action. Thus we have:

Rule 11:   Consequence -> {Reaction | Event}

Before turning to detailed examples in which this grammar has been applied to stories, it will be useful to review the various rules and terms of the grammar. Table I summarizes the entire set of syntactic and semantic rules. Table II consists of my definitions of the semantic relations used in the statement of the semantic interpretation rules and Table III gives verbal descriptions of the syntactic categories used in the formulation of the syntactic rules.

III. ANALYZING A STORY

I now turn to an example of how the grammar is to be used to discover the structure underlying a simple story. As an example, I will begin with the analysis of one of Aesop's Fables entitled "The Man and the Serpent". The text of the story is given in Table IV. It should be

TABLE III.
Definitions of Syntactic Categories

Action	An activity engaged in by an animate being or a natural force.
Application	The process of attempting to carry out some plan for obtaining a desire.
Attempt	The formulation of a plan and application of that plan for obtaining a desire.
Change-of-state	An event consisting only of some object changing from one state to another.
Consequence	The outcome of performing an action for the purpose of obtaining some particular state of the world.
Desire	An internal response in which one wants and will thus probably try to obtain some particular state of the world.
Emotion	An internal response which consists of the expression of feelings.
Episode	An event and the reaction of an animate being to that event.
Event	A change of state or action or the causing of a change of state or action.
Internal Response	The mental response of an animate being to an external event.
Overt Response	The willful reaction of a willful being to an internal response.
Plan	The creating of a subgoal which if achieved will accomplish a desired end.
Preaction	An activity which must be carried out in order to enable one to carry out a planned action.
Reaction	The repsonse of a willful being to a prior event.
Setting	The introduction to the characters and conditions of the characters in a story.
State	A property or condition of an object or a stable relationship among a set of objects.
Story	A kind of structured discourse which centers around the reactions of one or more protagonists to events in the story.
Subgoal	A goal developed in service of a higher more central goal.

TABLE IV.
The Text of "The Man and the Serpent"

(1) The Man and the Serpent
(2) A countryman's son, by accident, trod upon a serpent's tail.
(3) The serpent turned
(4) and bit him.
(5) He died.
(6) The father, in revenge,
(7) got his axe,
(8) pursued the serpent,
(9) and cut off part of his tail.
(10) So the serpent, in revenge,
(11) began stinging several of the farmer's cattle.
(12) This caused the farmer severe loss.
(13) Well, the farmer thought it best to make it up with the serpent.
(14) So he brought food and honey to the mouth of its lair,
(15) and said to it "Let's forget and forgive; perhaps you were right in trying to punish my son, and take vengeance on my cattle, but surely I was right to revenge him; now that we are both satisfied why should we not be friends again?"
(16) "No, no," said the serpent; "take away your gifts; you can never forget the death of your son, nor I the loss of my tail."

noted that the text is already parsed into lines containing single propositions. These lines will be treated as the units on which the grammar operates. This fable is somewhat unusual in that there is virtually no setting for the story. The introduction of characters is completed in the title and the story proper starts right out with an Episode. As we shall see shortly, this story is properly analyzed as a sequence of embedded episodes. To illustrate the analysis, I will begin with the first episode of the story. The first episode consists of lines (2)-(5) of the story. The syntactic structure of this episode is given in Fig. 2a and the semantic structure in Fig. 2b.

It should be noted especially that the internal response
and the Plan development both had to be inferred from
succeeding actions. As will be noted in our discussions of
summaries in the following section, it is very frequent that
the nature of the mental events are not expressed when
compressing the description of an event. Fig. 2b shows the
proposed semantic structure of this episode; the structure is
a completely right branching set of embedded causatives.
As we shall also discuss in the following section structures
such as these can be readily summarized by, for example,
simply stating the most deeply embedded causal structure.
In this case the structure (4) results in (5).

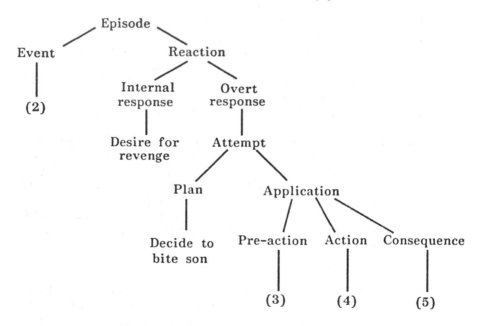

Fig. 2a. Syntactic structure of story.

Now, the next episode of the story actually includes this
first episode as the initiating event and lines (6) - (9) of
the story simply give us the farmer's Reaction to this
event. The structure of the second episode is essentially
the same as that for the first episode with the exception
that the Internal Response--the desire to get revenge--is

explicitly mentioned in the text. The second interesting aspect of the structure is that it seems that the action and the consequence are lexicalized together into the verb "cut off". This suggests that our grammar should really work on the level of decomposed predicates rather than surface propositions. On the other hand, the loss is not so great working with surface propositions and the process of applying the grammar is really much simplified.

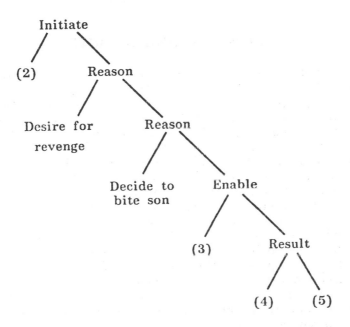

Fig. 2b. Semantic structure of story.

No new problems are introduced in the analysis of the remaining two episodes of the story. In each case, the prior episode serves as the initiating event for the immediately following episode.

In addition to this and the Margie story discussed above, the grammar has been applied in detail to four additional stories and applied informally to many others. The grammar, as described above, is adequate for all of the stories analyzed in detail and it is my impression that most folk tales, fables, and the like can be analyzed without many new problems arising.

Rather than continue with a detailed analysis of the grammar and how various stories can be analyzed in terms of it, I now turn to a discussion of what one can do with a grammar of the form outlined above. In the following two sections I discuss two distinct ways in which this grammar can be used to extend our understanding of the appropriate linguistic analysis of connected discourse.

IV. SUMMARIZING STORIES

Connected discourse differs from an unrelated string of sentences (among other ways) in that it is possible to pick out what is important in connected discourse and summarize it without seriously altering the meaning of the discourse. The same is not true of strings of unrelated sentences. Since such strings lack structure and do not make meaningful wholes they cannot be summarized at all.

In this section a set of rules are developed which map our semantic structures onto summaries of these structures. In order to evaluate the summary rules, eight subjects were asked to read each of six stories including the "Margie" story and "The Man and the Serpent". The subjects were asked to give brief summaries of each of the stories. [Subjects had copies of the stories in front of them while performing this task.] In the remainder of this section I will present and discuss the summarization rules developed and show how they relate to the summarization responses actually observed.

A. The Summarization Rules

Summarization rules are defined on each of the sorts of semantic structures that can be generated by our grammar. Thus there is one or more summarization rule for each of the substantive semantic relationships defined in Table II.

B. Summarizing CAUSEs

The relationship CAUSE[Event1,Event2] is summarized by according one of four different rules for summarizing such events. We will illustrate this with a set of examples Suppose, the following sentence is encountered:

(6) The snake bit the countryman's son causing him to die.

Here the CAUSE relationship holds between the event of the snake biting the boy and the event of his death. The following summaries are typical:

(7a) The snake caused the countryman's son to die.

or in a more naturally lexicalized formulation

(7b) The snake killed the countryman's son.

Here we have summarized the actual causal event by mere mention of the agentitive force in that causal event. Thus the following summarization rule:

S1: Summary(CAUSE[X,Y])->Agent(X) caused (Y)

It is frequently the case that the "cause" and the resultant (Y in this case) are lexicalized into one term (e.g., cause to die here becomes kill).

A second summarization rule applicable to the CAUSE relationship is much like the first; it involves the use of *instrument* rather than agent. Thus if we encounter the following sequence:

(8) The wind carried the balloon into the branch of a tree and it broke.

(8) would be interpreted as involving a CAUSE relationship between the two events in question. By rule S1, this can already be summarized as

(9) The wind caused the balloon to break.

or

(10) The wind broke the balloon.

Sentence (8) can also be summarized by

(11) The tree caused the balloon to break.

In this case the following rule was applied:

S2: Summary(CAUSE[X,Y])->Instrument(X) caused (Y).

Rules S1 and S2 together provide an account of how language summarizes experience as well as how one linguistic string summarizes another.

The third and fourth rules for summarizing CAUSEs are very similar and very common, but they are not as powerful and do not summarize as concisely.

S3: Summary(CAUSE[X,Y] -> (X) {0|and|so|caused} (Y)

Thus the proposition underlying (8) can be stated in any of four equivalent ways, and appears originally in form (c):

(12a) The wind carried the balloon into a branch of a tree. It broke.
(12b) The wind carried the balloon into a branch of a tree and it broke.
(12c) The wind carried the balloon into a branch of a tree so it broke.
(12d) The wind carried the balloon into a branch of a tree causing it to break.

The fourth summarization rule is:

S4: Summary(CAUSE[X,Y] -> (Y) {because|when|after} (X)

which puts the effect before the cause in three forms:

(13a) The balloon broke because the wind carried it into a tree.
(13b) The balloon broke when the wind carried it into a tree.

(13c) The balloon broke after the wind carried it into a
       tree.
(13d) The countryman's son died because a snake bit
       him.
(13e) The countryman's son died when a snake bit him.
(13f) The countryman's son died after a snake bit him.

Although our applications of rules S3 and S4 may not have
seemed to be very short summaries of (6) and (8), it should
be pointed out that the four events alluded to in (6) and
(8) are already summaries of the underlying events under
discussion.

C. Summarizing INITIATIONs

The highest node in an episode is an INITIATE
relationship between an Event and a Reaction to that event.
Thus the summarization of an episode is the summarization
of an INITIATE relationship. I have developed three basic
formula for the summarization of INITIATE relationships:

S5: Summary(INITIATE[X,Y]) -> (X)
     {∅ | and | caused | so} (Y)

S6: Summary(INITIATE[X,Y]) -> (Y)
     {because | when | after} (X)

S7: Summary(INITIATE[X,Y]) -> (Y)

Rules S5 and S6 are obvious analogs to Rules S3 and S4
and similar examples to those illustrated in (14) could be
given. Rule S7 illustrates the emphasis on the second
argument, the reaction, of the INITIATE relationship.
Thus Episodes can occasionally be summarized by a
summary of the Reaction alone with no reference to the
INITIATing Event. Summarization of this sort will be
illustrated somewhat later.

D. Summarizing ALLOWs

ALLOW(Event,Event) is a rather weak causal relationship with a strong emphasis on the ALLOWed Event. Thus summarizations of the ALLOW relationship do not include the explicit reference to causal terms and are very frequently summarized with reference only to the ALLOWed Event. We thus obtain the following summarization rules:

S8: Summary(ALLOW[X,Y] -> (X) {∅ | and} (Y)

S9: Summary(ALLOW[X,Y]) -> (Y) after (X)

S10: Summary(ALLOW[X,Y]) -> (Y)

In the serpent story, I analyzed the relationship between the snake turning and the snakes killing the boy to be an ALLOW relationship. This relationship can be summarized then, in the following ways:

(14a) The snake turned toward the boy. He killed him.
(14b) The snake turned toward the boy and killed him.
(14c) The snake killed the boy after he turned toward him.
(14d) The snake killed the boy.

E. Summarizing MOTIVATIONs

MOTIVATE is a relationship that holds between a thought and a response to that thought. It appears that MOTIVATEs are differentially summarized depending on the thought in question and depending on the degree to which the response to the thought adequately carried out the intentions of the thought. I thus formulated four summarization rules for MOTIVATE--one rule which applies generally, one for responses to Desires and two for responses to Plans.

S10': Summary(MOTIVATE[thought,response]) -> (response)

Rule 10 expresses the general tendency in summaries to delete mention of the thought and assume that the MOTIVATION of a response is fairly well predicted by the nature of the response. Thus in the Margie story there was no need to mention Margie's sadness at the breaking of her balloon because it was more or less implied by her crying. This is very common and perhaps in the bulk of the cases the "thoughts" are not explicitly referred to.

S11: Summary(MOTIVATE[Desire,response]) -> Desirer {got | managed to get | tried to get} (Desire) by (Response)

In this case the terms "got" or "managed to get" are used just in case the response in fact accomplishes the Desire. The phrase "tried to get" is used in the case the response failed to accomplish the Desire. Furthermore, the "by phrase" of the rule is optional and may be deleted. Thus we obtain summaries of portions of the serpent story such as:

(15) The man tried to make peace by bringing the snake food.

(16) . . . the snake got revenge by killing some of his cattle.

(17) . . . a farmer tried to make peace.

In the case where the thought is a Plan, a similar, but slightly different, rule seems to apply.

S12: Summary(MOTIVATE[Plan,response]) -> Planner {tried to do | did | managed to do} (Plan) {so | but | and | Ø} Consequent(response)

In this case again if the plan was unsuccessful, we obtain the phrase "tried to do plan" and "but Consequence" whereas if the plan was successful, we would be more likely to obtain "did plan", "managed to do plan" and "so consequence". One example from the serpent story is:

(18) The man managed to cut off part of the snake's
      tail.

(Note in this example the Consequence has been lexicalized
along with the action in the verb.)
      Finally, we have

S13: Summary(MOTIVATE[Plan,response]) ->
      Consequent(response) {when | because | after | by}
      Planner {tried to do | managed to do | did} (Plan)

Here we obtain examples such as:

(19) A dog lost his meat in a stream when he tried to
      grab its reflection in the water.
(20) A greedy dog lost his meat by trying to get the
      meat held by his reflection.

F. Summarizing ANDs and THENs

In addition to summarizing the above four semantic
relations, the two sequencing relationships are also
summarized. No formal rules will be attempted here, but
it can be said that ANDs and THENs are both summarized
primarily by deletion. Two rules seem to determine which
get deleted:

(1) Those arguments which dominate the fewest nodes
seem to get deleted first.
(2) In a temporally sequenced set of arguments the first
and the last are the last to be deleted. The last argument
in the sequence is usually the very last to be deleted.

G. Generation of Summaries

I will proceed, in this section, by showing how the rules
developed in the prior section can be used to account for
the summaries we have observed. I will begin with the

Margie story. Consider, first, the summary given by my first subject for this story.

(21) Margie cried when the wind broke her balloon.

I will now consider how an approximation to the summary (21) might be derived according to our summarization rules. With the story parsed according to the grammar, the summary for the entire story becomes just a summary of the episode through the deletion of the setting by rule S10. Sequential application of rules S6, S1, and S10 generate the following form of summary:

(22) Summary(reaction) *when* Agent(event1) *caused* Summary(event2)

We can then insert the propositional and nominal referents for the two summaries and the agent. Then if we delete the second of the conjoined arguments in "Margie cried and cried", and make the proposition embedded under cause into an infinitive, we obtain the summary:

(23) Margie cried when the wind caused her balloon to burst.

If we in addition realize that "burst" is "to become broken in a certain manner" and that "cause to become broken" is the same as "break" we can derive the same summary given by our subject.

Now consider the derivation of the summary for the serpent story:

(24) A farmer cut the tail off of a snake after it had killed his son. After the snake attacked the farmer's cattle, the farmer asked for a truce, but the snake refused.

The closest derivation of (24) is:

(25) A farmer cut off a serpent's tail after it killed his son. After the serpent stung the farmer's cattle, the farmer tried to make peace, but the snake refused.

These examples are designed to illustrate how the summarization rules can interact with the semantic and syntactic rules to produce the kinds of summaries we observe from our subjects. Obviously, a more detailed analysis is required to determine whether these rules will account for the summaries of many people over many stories. It does seem, however, that these rules are sufficient to account for the most obvious aspects of people's summarization over a number of stories.

V. CONCLUSION

The outline presented above is obviously only a tentative beginning of a theory of the structure and summarization of stories. The summarization rules are clearly incomplete and a good deal more work is required to apply them to more complex cases. Furthermore, the grammar has difficulty handling more complex multi-protagonist stories. Nevertheless, the results to date have been encouraging and generalizations to the rules listed above are currently under development. Moreover, a number of attempts to use these rules (and generalizations of these rules) as tools for the analysis of various kinds of experimental data are currently being made. For example, preliminary results indicate that "more structured stories" are easier to recall than less structured stories. An attempt is being made, within the grammar, to define the notion of degree of structure in a way consonant with the observed results. In a forthcoming paper I will report on a more detailed analysis of summarizations collected from additional subjects on a wider range of story types. A version of the grammar is under development which will account for the flow of a conversation and later summarization of that conversation. The structures generated by the grammar are being used to propose answers to *why-questions*. A multi-protagonist version of the grammar is under development. Although these results are not all in yet, it now appears that the basic character of the grammar presented above will allow us to account for a substantial range of phenomena related to the higher order structures found in stories.

Before concluding, a comment about the relationship of

this approach to that developed by other workers is in order. In this volume, the work most closely related to this is to be found in Chapters 9 and 10 by Schank and by Abelson. I see both of those lines of attack as generally consonant with my own. The work presented here differs from Schank's primarily in that I suggest global rules and attempt to describe the global structure of a story. Schank, on the other hand, emphasizes the "bottom-up" need to integrate each sentence into a paragraph. Using his methods, the first level relationships among the sentences become clear, but no systematic higher order structure is presupposed. It is my suspicion that any automatic "story parser" would require the "top-down" global structures that I propose, but would to a large degree discover them, in any given story, by the procedures developed by Schank.

With respect to work outside this volume it should be added that the approach developed here is, to some degree, designed to be a systematization of Propp's (1968) analysis of Russian folk tales. Although it hasn't been shown here, the rules presented above are designed to capture the relationships among the structures developed by Propp. In that light, this work should be compared to that of Colby (1973) on Eskimo folktales.

ACKNOWLEDGEMENTS

Research support was provided by grant NS 07454 from the National Institutes of Health and by grant GB 32235X from the National Science Foundation.

REFERENCES

Colby, B. N. A partial grammar of Eskimo folktales. *American Anthropologist*, 1973, 75, 645-662.

Norman, D. A., Rumelhart, D. E., and the LNR Research Group. *Explorations in cognition.* San Francisco: Freeman, 1975.

Propp, V. *Morphology of the folktale.* Austin: University of Texas Press, 1968.

Rumelhart, D. E., Lindsay, P. H., & Norman, D. A. A process

model for long-term memory. In E. Tulving & W. Donaldson (Eds.) *Organization of memory.* New York: Academic Press, 1972.

Schank, R. C. Causality and reasoning. Technical Report 1. Castagnola, Switzerland: Instituto per gli Studi Semantici e Cognitivi, 1974.

Schank, R. Identification of conceptualizations underlying natural language. In R. Schank and K. M. Colby (Eds.), *Computer Models of Thought and Language.* San Francisco: Freeman, 1973.

THE STRUCTURE OF EPISODES IN MEMORY

Roger C. Schank
Yale University
New Haven, Connecticut

I. INTRODUCTION

The past few years have significantly altered the direction of research in computer processing of natural language. For nearly the entire history of the field, parsing and generating have been the major preoccupations of researchers. The realization that meaning representation is a crucial issue that cannot be divorced from the above two problems, leads to a concern with the following two questions:

* How much information must be specified, and at what level, in a meaning representation?

* To what extent can problems of inference be simplified or clarified by the choice of meaning representation?

II. CONCEPTUAL DEPENDENCY

These are the principal issues we have been concerned with in our work on conceptual dependency: (1) It is useful to restrict severely the concepts of action such that actions are separated from the states that result from those actions. (2) Missing information can be as useful as given information. It is thus the responsibility of the meaning representation to provide a formalism with requirements on items which must be associated with a concept. If a slot for an item is not filled, requirements for the slot are useful for making predictions about yet to be received information. (3) Depending on the purposes of the understanding processes, inferences must be made that will (a) fill in slots that are left empty after a sentence is completed (b) tie together single actor-action complexes (called "conceptualizations") with other such complexes in order to provide higher-level structures.

Throughout our research it was our goal to solve the paraphrase problem. We wanted our theory to explain how sentences which were constructed differently lexically could be identical in meaning. To do this we used the consequence of (1) and (2) to derive a theory of primitive ACTs. Simply stated, the theory of primitive ACTs states that within a well-defined meaning representation it is possible to use as few as eleven ACTs as building blocks which can combine with a larger number of states to represent the verbs and abstract nouns in a language. For more detail on these acts, the reader is referred to the Appendix and to Schank (1973a) and Schank (1975). We claim that no information is lost using these ACTs to represent actions. The advantage of such a system is this:

(1) paraphrase relations are made clearer,

(2) similarity relations are made clearer,

(3) inferences that are true of various classes of verbs can be treated as coming from the individual ACTs. All verbs map into a combination of ACTs and states. The inferences come from the ACTs and states rather than from words.

(4) Organization in memory is simplified because much information need not be duplicated. The primitive ACTs provide focal points under which information is organized.

A. Some Caveats about Oversimplification

The simplest view of a system which uses primitive ACTs gives rise to a number of misconceptions. The following caveats, which we subscribe to, must be observed.

(1) There is no right number of ACTs. It would be possible to map all of language into combinations of mental and physical MOVE. This would, however, be extremely cumbersome to deal with in a computer system. A larger set (several hundred) would overlap tremendously causing problems in paraphrase recognition and inference organization. The set we have chosen is small enough not to cause these problems without being too small. Other sets on the same order of magnitude might do just as well.

(2) The primitive ACTs overlap. That is, some ACTs nearly always imply others either as results or as instruments. Our criteria for selecting a given ACT is that it must have inferences which are unique and separate from the set of already existing ACTs.

3) Information can be organized under each primitive ACT, but it is definitely necessary to organize information around certain standard combinations of these ACTs. Such "super predicates" should be far fewer than those presented by other researchers, but they certainly must exist. Many inferences come from "kiss" for example that are in addition to those for MOVE. Organizing information under superpredicates such as "prevent" is, however, misguided since prevent is no more than the sum of its individual parts.

(4) Information is not lost by the use of primitive

ACTs, nor is operating with them cumbersome in a computer program. Goldman (in Schank, 1975) has shown that it is possible to read combinations of ACTs into words very easily. Rieger (1974) has written a program to do inferences based on a memory that uses only those ACTs. Riesbeck (1974) has shown that predictions made from the ACTs can be used as the basis of a parsing program that can bypass syntactic analysis. The MARGIE program (see Schank, Goldman, Rieger, & Riesbeck, 1973) would seem to indicate that objections to ACTs on computational grounds are misguided at the current level of technology.

B. Toward Episode Representation

This chapter outlines our assumptions about how people tie together episodes or stories. We use the notions of conceptual dependency and primitive ACTs to represent and paraphrase entire episodes. We assume that knowing how to build larger structures bears upon the problem of how to use context to direct inferencing, and when to stop making inferences. The program of Goldman and Riebeck (1973) paraphrased sentences semantically but did not go beyond the level of the sentence. The solution that we will present is meant to be a partial solution to the paraphrase problem.

Our goal is to combine input sentences which are part of a paragraph into one or more connected structures which represent the meaning of the paragraph as a whole. While doing this we want to bear in mind the following: (1) A good paraphrase of a paragraph may be longer or shorter (in terms of the number of sentences in it) than the original. (2) Humans have little trouble picking out the main "theme" of the paragraph. (3) Knowing what is non-essential or readily inferable from the sentences of a paragraph is crucial in paraphrasing it as well as parsing it.

We pose the question of exactly how units larger than the sentence are understood. That is, what would the resulting structure in intermediate memory look like after a six-sentence episode or story had been input? Obviously, there must be more stored than just the conceptual dependency representation for each of the six sentences, but we could not store all possible inferences for each of the six sentences. Ideally what we want is to store the six sentences in terms of the conceptualizations that underlie them as an interconnected chain. That is, we shall claim that the amount of inferencing which is useful to represent the meaning of a paragraph of six sentences in length is precisely as much as will allow for the creation of a *causal chain* between the original conceptualizations.

Finally we shall address the question of creating the best model to explain how information is stored in human memory. We do not believe that people remember everything that they hear, but rather that forgetting can be explained partially in terms of what information is crucial to a text as a whole and what information can be easily rediscovered.

C. Causal Links

We shall claim here that the causal chain is the means for connecting the conceptualizations underlying sentences in a text. Conceptual dependency allows for four kinds of causal links. These are:

Result Causation: An action causes a state change. The potential results for a given action can be enumerated according to the nature of that action. Consider some illustrative examples: *John hurt Mary* - John did something which *resulted in* Mary suffering a negative physical state change. *John chopped the wood* - John propelled something into wood which resulted in the wood being in pieces.

Enable Causation: A state allows for an action to have the potential of taking place. Here too, the states necessary for an action to occur can be enumerated. *John read a book* - John's having access to the book and eyes, etc. *enabled* John to read. *John helped Mary hit Bill* - John did something which *resulted* in a state which enabled Mary to hit Bill. *John prevented Bill from leaving* - John did something which ended the conditions which enabled Bill to go. (This is *unenable* causation.)

Initiation Causation: Any act or state change can cause an individual to think about (MBUILD) that or any other event. *John reminds me of Peter* - when I think of John it initiates a thought of Peter. *When John heard the footsteps he got scared* - hearing the footsteps *initiated* a thought about some harm that might befall John.

Reason Causation: An MBUILD of a new thought can usually serve as the reason for an action. Reason causation is the interface between mental decisions and their physical effects. *John hit Mary because he hated her* - John thought about his feelings about Mary which made him decide to hit her which was the *reason* he hit her.

In this chapter causality will be denoted by a line from cause (top) to effect (bottom). A label will indicate which causal it is. Sometimes two or more labels will appear at once, which indicates that all the intermediate conceptualizations exist, but that we simply choose not to write them.

III. UNDERSTANDING PARAGRAPHS

Consider the following three paragraphs constructed from three different introductions (S1, S2 and S3) followed by a paragraph (BP). The base paragraph is:

(BP) John began to mow his lawn. Suddenly his toe started bleeding. He turned off the motor and went inside to get a bandage. When he cleaned off his foot, he discovered that he had stepped in tomato sauce.

(S1) It was a warm June day. (followed by) BP.

(S2) It was a cold December day. (followed by) BP.

(S3) John was eating a pizza outside on his lawn. He noticed the grass was very long so he got out his lawn mower. (followed by) BP.

We will refer to S1 (S2, S3) followed by BP as paragraph I (II, III).

Within the context of a paragraph, a sentence has a dual role. It has the usual role of imparting information or giving a meaning. In addition, it serves to set up the conditions by which sentences that follow it in the paragraph can be coherent. Thus in general, in order for sentence Y to follow sentence X and be understood in a paragraph, the conditions for Y must have been set up. Often these conditions will have been set up by X, but this is by no means necessarily the case. The conditions for Y might be generally understood, that is, part of everyday knowledge, so that they need not be set up at all. If this is not the case then it is irrelevant that X precedes Y.

A. Necessary Conditions

By a *necessary condition*, we mean a state (in conceptual dependency terms) that enables an ACT. Thus one cannot play baseball unless one has access to a ball, a bat, a field to play on, etc. The conditions that are necessary in order to do a given ACT must be present before that ACT can occur.

It is thus necessary, in order to understand that a given ACT has occurred, to satisfy oneself that its necessary conditions have been met. Often this requires finding a causal chain that would lead to that condition being present. That is, if we cannot establish that John has a bat, but we can establish that he had some money and was in a department store, we can infer that he bought one there if we know that he has one now. Obviously there are many possible ways to establish the validity of a necessary condition. Often, it is all right just to assume that it is "normally" the case that this condition holds. Rieger (1974) makes extensive use of this kind of assumption in his inference program. Establishing or proving necessary conditions is an important part of tying diverse sentences together in a story (John wanted to play baseball on Saturday. He went to the department store.). It is also, as we shall see, an important part of knowing when a paragraph does not make sense.

Thus the problem of representing a paragraph conceptually is at least in part the problem of tying together the conditions set up by sentences with the sentences that required those conditions to be set up. In the base paragraph (BP) no conditions have been set up under which tomato sauce could reasonably be considered to be present. Thus BP does not hang together very well as a paragraph. With S3, however, the conditions for tomato sauce being present are set up, the last sentence of BP can be tied to those conditions, and this is an integral paragraph. Note that there is no need to set up "bandage's" existence, since people can normally be assumed to have a bandage in a medicine cabinet in a bathroom in their house. If the third sentence were replaced by: "He took a bandage off of his lawn mower," however, there would be problems.

What we are seeking to explain then is what the entire conceptual representation of a paragraph must be. It can be seen from the above that in our representation of BP, it might well be argued that "bathroom" and "medicine cabinet" are rightfully part of the representation. The representation of a paragraph is a combination of the conceptualizations underlying the individual sentences of the paragraph plus the inferences about the necessary conditions that tie one conceptualization to another or to a given normality condition.

Let us consider the connectivity relationships with S1. We shall do this by looking at the necessary conditions required for each conceptualization and established by that conceptualization for subsequent conceptualizations.

S1, *It was a warm June day*, consists of two details. One is that the time of some unspecified conceptualization was a day in June. The other is that the temperature for that day was warm (or to be more precise greater than the norm for that season). These two states enable all warm weather activities. That is: the *established conditions* invalidate any activities for which contradictory conditions would be necessary.

Consider as an example S1 followed by an S4 such as: S2. John began to build a snowman. The fact that people would find such sequences bothersome, if not absurd, indicates that part of the process of understanding is the tying together of *necessary conditions* and *established conditions*. Necessary conditions are backward looking inferences that must be generated for each input conceptualization. For S4 we would have something like -- "check to see if the necessary condition for this action (coldness) has been established. If it has not been, then infer it unless a contradictory condition has been established in which case there is an anomaly."

Established conditions are forward-looking inferences. These are states which are inferable from an input conceptualization. Often established conditions are input directly and need not be inferred at all. Thus S1 provides two established conditions. Neither of these conditions are called into play until a request for their existence is received from a following input conceptualization.

Now the actual S2 in paragraph I is "John began to mow his lawn". S2 requires appropriate necessary conditions to be satisfied. Some of these are: (a) John has a lawn; (b) John has a lawn mower; (c) John's lawn has grown to a length such that it might be mowed; (d) It is a pleasant day for being outside. (Obviously, there are an extremely large set of conditions to be satisfied. Part of the problem is knowing how to reduce the search space.)

It can be seen from these four conditions that they are not all equal. That is (a) and (b) are of the class of necessary conditions that we shall call "absolutely necessary conditions" (ANCs). An ANC is a state which must be present in order to enable a given ACT to take place. If an ANC is violated, a sentence sequence is construed to be anomalous. Conditions (c) and (d), however, are the more interesting. They are what we shall call "reasonable necessary conditions" (RNCs). An RNC is a state which is usually a prerequisite for a given ACT. If an RNC is violated, the story sequence is not interrupted. Rather a peculiarity marking is made. These peculiarity markings turn out to be a crucial part of story understanding because they are predictive. That is, the sequence "It was a cold December day. John walked outside in his bathing suit," predicts a to-be-related consequence which is likely the point of the story. What makes this prediction is the peculiarity marking that would be generated from the established violation of the RNC involving warm weather necessary for walking outside scantily clad.

Necessary conditions are thus a subset of the set of possible inferences. They are precisely the set which is commonly used to connect isolated conceptualizations together. Necessary conditions are checked by means of establishing whether there exist records of them in memory. First, there is an attempt to tie them to information already present from previous sentences. This is done by use of the general facts in memory (e.g., about warm weather and lawns) including normality conditions. If such facts are present or can be established, an enable causation link is established between the previously input or inferred state and the new input conceptualization. That is, it is quite usual for people to possess lawns and lawnmowers so the fact that nobody told us this about John is not upsetting. (Of course, if we knew that John lived in an apartment then an ANC would be violated and there would be trouble.) If the connection can be established from normality information, then the new states involving this normality information are inferred as being the case in this particular instance (i.e., John has a lawn is inferred). When none of this can be done, a severe problem in the cohesion of the story (for the listener) exists.

The interesting part of paragraph connectivity comes when we examine the sentence "Suddenly his toe started bleeding." How can we connect this sentence to previous information? In fact humans do it with little trouble. A question is generated, "what can cause bleeding in people?" This question is coupled with an inquiry as to whether the established conditions demandable from any previous information in the text result in "a human bleeding" under the right circumstances. Working in both directions at once, we can establish a chain from "lawn mowing" to "blades turning" to "toe bleeding" if we hypothesize contact between blade and toe. The syntax for causality in this chain follows exactly the same lines as proposed by Schank (1973b). The kind of inference outlined previously has been considered in some detail (including programming some examples on the order of complexity shown here) by Rieger (1974). Let us now consider what the final representation of paragraph I should be.

(In our representation we shall use a simplified

conceptual dependency diagram, consisting of a parenthesized expression in the order: (ACTOR ACTION OBJECT X) where X is any other relevant case which we choose to include which we make explicit there. State diagrams are written: (OBJECT NEWSTATE). Necessary conditions are abbreviated as ANCs or RNCs. Ss designate input sentences. INFs designate inferences needed to complete causal chains. The time of an event or state is relative to the others around it with earlier events higher on the page. story (for the listener) exists. Fig. 1 is the final representation of the paragraph except for the following:

(1) Normally the initiation reason causation (from ANC/3) would have to be expanded. In this case there would have to be a belief about what bandages had to do with bleeding body parts. This belief would be part of the MBUILD structure that supplied the reason for going in the house.

(2) All the ANCs would have to be accounted for. This would be done as the paragraph was being processed.

The ANCs for the above paragraph are as follows:

ANC/1, ANC/2,...:   These specify that John has a lawn and a lawn mower and can push it, etc.

ANC/3:   John has a toe and he was not wearing shoes, etc.

ANC/4:   John has a house, it is nearby and he can get in, etc.

ANC/5:   John has a bandage in his house, and he can find and get it.

ANC/6:   This is provided by "John cleaned his toe". Thus the entire structure causally related to ANC/6 established another chain apart from the actual story.

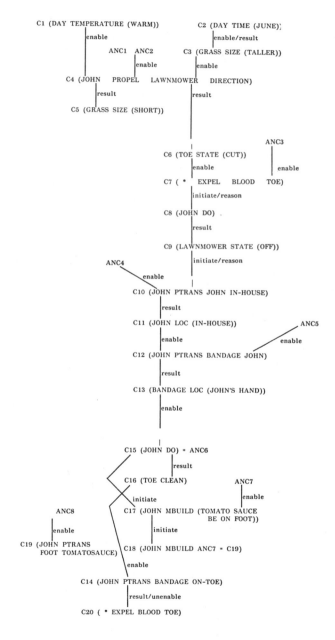

Fig. 1. Conceptual dependency representation of the paragraph.

ANC/7:   This is specified by the next part of the last sentence.

ANC/8:   This is the condition that tomato sauce be in a place where John might have stepped in it.

Two important things are illustrated in this diagram of paragraph I: (1) Conceptually, a paragraph is essentially a set of causal chains, some leading to dead ends and at least one carrying on the theme and point of the story. (2) As long as required necessary conditions can be established and inferences necessary to complete causal chains can be resolved, a paragraph is coherent and understandable. When these processes are too difficult or impossible problems result.

In paragraph I, ANCs 1-5 are easily satisfied by everyday normality conditions, but ANCs 6-8 are more of a problem. ANC/6 is provided by the story in C15, but C17 which is initiated by ANC/6 requires John himself to establish (in an MBUILD) ANC/7. In trying to establish ANC/7 John concludes that C19 is true. This leaves the reader with the problem of verifying ANC/8, which is not possible under these conditions. The rest of the story thus remains incomplete and can never really get to C14 and C20.

Now, in paragraph III on the other hand, ANC/8, when it is discovered, is easily resolvable. (PIZZA LOC (LAWN)) is a possible result of the first sentence (John was eating a pizza outside of his lawn). If the problem of knowing that a pizza contains tomato sauce is resolvable, then so is ANC/8. In that case, an ANC would be resolved from within the given paragraph as was done for ANC/6 and ANC/7 in paragraph I.

Some additional comments that can be made about the representation of paragraph I are:

(1) C5 is a deadend. Short grass may be the condition for something, but this story does not tell us nor make use of it. Deadend paths in a story indicate items of less importance in the story. This information is crucial in the problem of paraphrase of paragraphs since it tells you what you can leave out.

(2) The inference of the result link between C4 and C6 is one of the most important inferences in the story. It serves to tie together two apparently unrelated events into a contiguous whole.

(3) The information about turning the lawnmower off is another deadend. In fact, there was no reason for this information to be in the story at all.

(4) C14 and C20 are crucial to a story that has not been told. What has happened is that the attempted resolution of ANC/6 has interrupted the flow of the story. Inside the resolution of ANC/6 we have had to resolve ANC/8 which causes us to quit.

Diagrammatically we can view the story as in Fig. 2 which shows a version of the story structure with only concept labels. From various sources of information we can construct a path that sets up C4. One of the inferences from C4 is fruitful in that it provides a chain to C7. Two paths come from C7, one is fruitful in leading us to C10 (whose conditions can be explained by ANC/4). On our way to C14, we try to establish ANC/6. Doing so leads to a direct path to C18. In order to prove the conditions for C18, however, we need ANC/8 which cannot be determined.

B. Remembering Paragraphs

We have established in the previous section some basic principles regarding what we would expect to be the result in memory after a paragraph has been input. (1) The conceptual dependency representation of each input sentence is included. (2) The conceptualizations that underlie the sentences should connect to each other conceptually. (3) The basic means of connecting the conceptualizations underlying the input sentences to each other is the causal chain. (4) Inferences from the input conceptualizations are part of the representation of the total paragraph if they are used in order to connect input conceptualizations into a causal chain. (5) The necessary conditions must be satisfied for every represented conceptualization. This is done by inferring facts both from inside and outside the paragraph. Inferences made from outside the paragraph

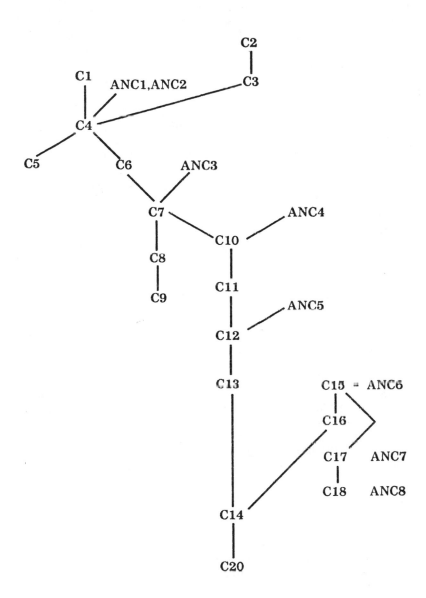

Fig. 2. A diagrammatic version of the paragraph.

proper are still part of the representation of the total paragraph. (6) Stories can be viewed as the joining together of various causal chains that culminate in the "point" of the story. Deadend paths that lead away from the main flow of the story can thus be considered to be of lesser importance.

From the point of view of paraphrasing tasks or the problem of remembering, the things most likely to be left out in a recall task are: the deadend paths; the easily satisfied necessary conditions, whether they were explicit in the original paragraph or not; and the inferences that make up the causal chain that are "obvious" and easy to recover at any time.

In order to better examine the validity of our predictions about memory and the usefulness of our representation for stories, we diagrammed the well known "The War of the Ghosts", a story used by Bartlett (1932) for experiments in memory. We are not interested here in all the facets of memory that Bartlett considered, but some of the problems that he concerned himself with can be handled more easily using our representation. The story is:

> One night two young men from Egulac went down to the river to hunt seals, and while they were there it became foggy and calm. Then they heard war-cries, and they thought: "Maybe this is a war-party". They escaped to the shore, and hid behind a log. Now canoes came up, and they heard the noise of paddles, and saw one canoe coming up to them. There were five men in the canoe, and they said:
>
> "What do you think? We wish to take you along. We are going up the river to make war on the people".
>
> One of the young men said: "I have no arrows".
>
> "Arrows are in the canoe", they said.
>
> "I will not go along. I might be killed. My relatives do not know where I have gone. But you", he said, turning to the other, "may go with them."
>
> So one of the young men went, but the other returned home.

And the warriors went on up the river to a town on the other side of Kalama. The people came down to the water, and they began to fight, and many were killed. But presently the young man heard one of the warriors say: "Quick, let us go home: that Indian has been hit". Now he thought: "Oh, they are ghosts". He did not feel sick, but they said he had been shot.

So the canoes went back to Egulac, and the young man went ashore to his house, and made a fire. And he told everybody and said: "Behold I accompanied the ghosts, and we went to fight. Many of our fellows were killed, and many of those who attacked us were killed. They said I was hit, and I did not feel sick".

He told it all, and then he became quiet. When the sun rose he fell down. Something black came out of his mouth. His face became contorted. The people jumped up and cried.

He was dead.

Using a diagram for this story based on causal chains, we followed the rules given below to prune the causal chains to create a summary diagram which would allow a paraphrase to be generated on later remembering. We have not worked out a complete set of rules, but the following give a reasonable first approximation:

(1) Deadend chains will be forgotten.
(2) Sequential flows (correct chains) may be shortened.
    (a) The first link in the chain is most important.
    (b) Resolution of questions or problems is too.
(3) Disconnected pieces will be either connected correctly or forgotten.
(4) Pieces that have many connections are crucial.

Using a diagram with only the causal connections (as shown in Fig. 2 for paragraph I) and the above rules, we derived a paraphrase which was:

Two men went to the river. While they were there they heard some noises and hid. Some men

approached them in canoes. They asked them if
they would go on a war party with them. One man
went. He got shot. The men took him home. He
told the people what happened and then he died.

The reasonableness of this paraphrase suggests that the above procedural outline for paragraph paraphrase is a good one.

If one compares our paraphrase with the output of Bartlett's subjects, the one striking difference is that the two conceptualizations missing necessary conditions are present in his subjects' output. We hypothesized that peculiarity markings would be generated for violation of necessary conditions. C37 and C52 ("They are ghosts," and "Something black came out of his mouth.") would have generated peculiarity markings. If Bartlett's output is examined, it can be seen that his subjects handled these peculiarity markings in various ways. They were made part of the causal chain in entirely different ways for each speaker and were hardly ever forgotten.

Paraphrases can be generated from meaning representations of text by procedures that read out the conceptualizations which are central to the flow of the diagram. Paraphrases longer than the original text would be generated by realizing all the conceptualizations in the final meaning representation. Paraphrases shorter than the text would be generated by various means. Among these are: (a) leaving out deadend paths, (b) only realizing conceptualizations that have more than two pointers to them in the text, (c) reading out only starting and ending points of the subchains in a text. Summaries would be developed in similar ways.

The main problem in building a computer program to do paraphrases is, at this point, not the generation aspect, but the problem of actually being able to make the crucial inferences that connect texts together. In the base paragraph the inference that the lawnmower may have cut John's toe is difficult to make. In "The War of the Ghosts" the crucial inference that the notion of ghosts somehow resolves the unconnectable causal chain at the end of the story is too difficult for humans to figure out.

IV. LONG-TERM MEMORY

Text is not really remembered verbatim. At some point there must be an integration of new input information within the long-term store. We have outlined some forgetting heuristics above, but they are only part of the issue. We must investigate as well, what we imagine long-term memory to look like, or, with respect to what we have just done here, how would the conceptual content in "The War of the Ghosts" be stored in long-term memory?

We shall advance the view here that a story is stored as a whole (albeit partially forgotten) unit, not decomposed. Such a view claims that any sequence of new information forms an episode in memory and any piece of information within an episode can be accessed only by referencing the episode in which it occurs. This view is similar to the view of memory expressed by using frames (e.g., Chapter 7). We will contrast this with the view that most information is stored in a hierarchically organized semantic memory.

As originally proposed by Quillian (1968), semantic memory was an associative network of words which were intended to represent the meaning of sentences as well as the knowledge used in understanding sentences. More recent modifications to Quillian's ideas on semantic memory have suggested network models which were both better structured in terms of the kinds of associative links allowed as well as more representative of the meaning of sentences. For example, Rumelhart, Lindsay, & Norman (1972) and Anderson & Bower (1973) each proposed a semantic memory which relates words together, and this same semantic memory was intended to account for meaning and world knowledge. Tulving (1972) proposed that this memory ought to be divided into two distinct pieces: a hierarchical portion containing static knowledge about relations between "words, concepts, and classification of concepts"; and an episodic portion which contains information gained through personal experience.

What we shall argue in the remainder of this chapter is that the distinction between semantic memory and episodic memory is a false one. We shall argue that what must be present is a lexical memory which contains all of the information about words, idioms, common expressions etc, and which links these to nodes in a conceptual memory, which is language free. We believe that it is semantic memory rather than episodic which is the misleading notion. Once we change semantic memory by separating out lexical memory, we are left with a set of associations and other relations between concepts that could only have been acquired by personal experience. We claim that conceptual memory, therefore, is episodic in nature.

A. Supersets in Memory

The case of much of semantic memory work relies on the notion of superset to organize the lexical data. Is there any reason to believe that humans really use supersets in their memories? To some extent the Collins and Quillian (1969) experiments indicate that some supersets are used by people. But it is important to point out where they are probably not used.

The 50 States Problem: A classic barroom bet involves asking someone to name all 50 states in the U.S. Very few people can actually name all 50 of them (including professors). It is not that they do not really "know" all 50. That is, they can recognize each of them as well as produce the names of each of them in response to specific queries (e.g., What state is Baltimore in?), but in the task of naming all 50, people find that their method for organizing the data is not a simple list called "States of the U.S." under which all 50 are stored. Some people try to visualize a map and read off the names. Others associate specific experiences with each.

The point is this: The information about states is organized around experiences. People probably have a marker after their memory token for a state saying that it is a state. They have no effective procedure that can "get all the tokens that have state on their property list." Rather, they try to find other pieces of their memory that have pointers to the tokens for the states (i.e., "places I went on my trip in 1971"). Not all the states can be located in this fashion and they are left out. Conrad's (1972) experiments suggest that such information is not necessarily stored in the most economical fashion possible. In other words, while general information exists, it is not part of the actual day to day working part of memory. Thus certain superset information may well exist in memory, but as part of an infrequently used general knowledge store, rather than as part of the core of working memory.

The Rabbit Problem: Suppose you wanted to code in your memory that "Rabbits eat carrots." A hierarchical organization suggests that carrots should be listed under the node "food". It is plain, however that carrots are only food for some animals, not all. The best organization would be one where experience codes static information. That is, "carrot is a potential food for rabbits" is more optimally coded with what we might call a "FOOD" link. That is, there is an associative link from carrot to rabbit. This link has a name which we can write as FOOD.
We are introducing here two concepts: (1) Links can themselves be linked to conceptualizations that describe the nature of the link. There are as many types of associations between concepts as there are conceptualizations in which they could possibly occur. (2) Hierarchical superset nodes are eliminated for functional supersets like food since information about food would be bound under particular instances of animals eating what is good for them. (This says nothing about nonfunctional supersets such as "bird".)

Contextual Organization: Naive organization of superset relationships fails to group contextually related entities. Thus while hammer, saw, and plunger all are tools, the first two are certainly more closely related to each other than they are to the third. This can be handled by having subheadings in the hierarchical tree, but this merely worsens an already mistaken notion. Concepts which have no place under an organizing node such as "tool" would still be closer to some tools than some tools would be to other tools. Thus we propose that "wood" is closer to "saw" than "plunger" is to "saw", even though the latter two are both tools. Hammers and nails and saws and wood only relate to each other through the conceptualization that involves their function and is part of various episodes. Thus "tool" is not a category name, but rather the specific instance of a tool is its relationship to an episode of its use.

In memory I propose that we might have an abstract conceptualization C1 with variables O1, O2, D1, and D2:

```
C1:  ACTOR:        ONE
     ACTION:       PROPEL
     OBJECT:       O1
     DIRECTION:    D1
     INSTRUMENT:   ACTOR:        ONE
                   ACTION:       PROPEL
                   OBJECT:       O2
                   DIRECTION:    D2
```

In episodic memory, a particular instance of C1 might contain O1=NAIL, D1=WOOD, O2=HAMMER, and D2=NAIL to describe a nail propelled toward wood by the act of propelling a hammer toward the nail. Other less filled instances of C1 might contain only

```
O1=NAIL      O2=HAMMER      (Hammering a nail)
D1=WOOD      O2=HAMMER      (Hammering wood)
```

Another instance of C1 might contain

```
D1=WOOD      O2=SAW         (Sawing wood)
```

There is no conceptualization containing both hammer and plunger. Thus the relationship between these objects is not direct; both occur in instances of C1 as object O2. I predict that finding these relations by analogy is a more difficult (and hence time-consuming) process than finding relations within a single instance of a conceptualization. I believe that the *definition* of a nominal concept (e.g., "tool") is a functional description of its use within an episode in memory (e.g., O2 in C1). Hammer and saw are closer than hammer and plunger because the former share a common element, D1=WOOD, in instances of C1.

Nonstatic Context: Going on with our discussion of what is likely to be close to what in a memory, we would claim that it is shorter from goulash to chicken paprikash than it is from goulash to spaghetti. This is a general statement that could be wrong for any particular individual. That is, someone who ate goulash only once and at the time had the best spaghetti of his life is likely to have a different organization.

For our average person we are again faced with the seemingly simplest choice of having a subset entitled "Hungarian food". The enticing properties of this are that things which are like each other would be grouped together. We would claim, however, that "the time I was in Budapest" and "the gypsy band that played while I ate goulash" are still closer to goulash than goulash is to spaghetti (where closeness is defined with respect to the number of links separating two concepts).

We are claiming, then, that in an actual memory, experiential groupings are the core of organization. supersets give way in the memory to organizing contexts and, once a context is entered, a person is more likely to think of another item in that context than an item from a different context that might be classed in the same superset.

B. Associative Links

We claim that the basis of human memory is the conceptualization. Internally the conceptualization is action-based with certain specified associative links between actions and objects. Externally, conceptualizations can relate to other conceptualizations within a context or episodic sequence. Objects are related to other objects only by the internal or external associations within the conceptualizations of which they are both part. That is, objects cannot be separated from the action sequences in which they are encountered. Objects relate to the episodes in which they occur, which may include other objects.

We classify the links in a memory to be of two kinds, intercontextual and intracontextual. The intracontextual links have as a subset all the links of conceptual dependency. Since the elements that make up a context are episodes which are coded into conceptualizations, this follows directly.

We define a context as any group of links which do not include intercontextual links. The intercontextual links are:

(1) *ISA*: superset
(2) *PROPS*: shares properties with.

An example of a natural path within a context would be, "Going to the museum in Berlin reminded me of going to the museum in Boston." An example of a jump between contexts would be "The museum in Berlin reminded me of an old hotel."

In the former example it is important to point out that the relation is between the *events* of going to the museum, not the museums themselves. Thus a memory should have in it the general paradigm of "museum going" and relate together the two instantiations of that paradigm simply because they both are connected to that paradigm. Thus the two particular events are related to each other directly through the paradigm of the event that they each represent.

In the second example, the museum only reminds one of the hotel because of shared properties. This represents a shift in contexts and is in fact likely to get one thinking about something that happened in that hotel which would be quite unrelated to the museum. This is where the *PROPS* link comes in. A context can be left (the museum) by going to another context which is related to it be a set of shared properties.

C. Rules of Association

What constitutes a reasonable path in memory? People usually agree that (A) is reasonable:

(A) First I thought of a pen, and then I thought of the paper I was writing.

A deviant path would be (B):

(B) First I thought of a watermelon, and then I thought of the paper I was writing.

What are the rules for reasonable paths? The last example is, of course, an unreasonable path as long as there are no intermediate steps to be filled in. We could argue that there must be a certain kind of path that a person follows to get from "watermelon" to "paper". For example, if there were watermelon stains on his paper, or the paper was about watermelon production, or first he ate a watermelon and then he wrote the paper, these would all be considered reasonable paths.

We wish to do more than just be able to point out when a statement seems well structured. We are concerned with finding the rules which allow a person to identify unacceptable structures and fill in for himself what he considers possible explanations of what the speaker "must have meant". Guessing the speaker's intent is of less importance than being able to ask a speaker for clarification at a particular point in a conversation.

With respect to causality we have listed the specific causal rules that allow one to fill in gaps in causality statements. We would claim that memory is structured using the same causal rules as stories. To this it is necessary to add rules which allow for tying together of statements that are not causally related. Consider the following sentence:

Seeing my mother reminded me of ice cream.

It is a usual inference to assume that this person had, according to some prescribed rules of concept organization, a relatively short path between these two concepts. People, in hearing the above sentence, can make guesses about what that path might be, for example:

(a) perhaps his mother used to make ice cream for him;
(b) perhaps he and his mother used to go to an ice cream store together;
(c) perhaps his mother told him to stop eating so much ice cream;
(d) perhaps his mother worked in an ice cream factory.

What is important about the above guesses is that to most people they would account for the path between the two concepts. That is, people can recognize (as well as generate) possible conceptualizations which relate two concepts. They do this in order to make sense of relatedness statements, or more importantly, to know when some idea "follows" from another. This latter is important in listening to a discourse in order to know that it is necessary to infer a connection between thoughts that were not explicitly stated or to question the lack of cohesion of a string of ideas. We establish a set of rules for this linking. The first rule is:

I.   Conceptualizations connect together the individual concepts that make them up.

That is, in order to get from one concept to another in memory, there must be a conceptualization of which they are both a part. The above rule is incomplete in that there is perhaps a more important corollary to it:

II.   If two concepts do not exist in one conceptualization in memory, there must exist a causal chain between two separate conceptualizations in which they occur in order to associate them in memory.

What we are doing, then, is defining the notion *association*. We are establishing that an associative conceptual memory is made up of the links that exist within a conceptualization. Tokens of a concept type can lead to tokens of another concept type by means of the conceptualizations that connect them. Such linkages explain associations and eliminate the need for supersets and other artificial relationships.

There is one further corollary :

III.   Two concepts can be related by their occurrence in separate conceptualizations which are part of the same episode in memory.

These three rules indicate that an association in memory must be initiated by conceptualizations which are part of episodes. Thus conceptual memory can be considered to be a series of conceptualizations linked together causally or temporally in episodic sequences.

D. Discussion

We have been arguing for a combination of the notions of semantic memory and episodic memory. In particular, we have said that all semantic information is encoded in episodes in the memory. We have further argued that the notion of superset in memory is vastly over-used and that much superset information can be better stored as episodes.

Although the Collins and Quillian (1969) experiments have met with some dispute, we feel that the basic idea is correct. That is, some supersets must be used which store the features of the class. This is done in order to save space, but more importantly, it is necessary for information which was derived from sources other than direct experience. If you have never seen an orangutang, it still is easy to know that it must breathe because we know it is an animal.

We would propose that very few supersets (maybe no more than 10) would actually be used in the memory as nodes under which information is stored. Thus "food" above has a definition but is only a lexical entry, not really a node in memory. This is because no information is actually stored under "food" per se. Likewise, "mammal" is only a dictionary entry which calls up the superset "animal" and makes a feature distinction within that class. This explains why people have trouble remembering if a whale is a fish or a mammal. A whale would be considered to be a fish in our memory, but one that might have information stored with it that related it to the feature change produced by "mammal" under "animal".

Thus we hold the view that semantic memory really is a misnomer and furthermore that the distinction between semantic memory and episodic memory is wrong. Once lexical memory is separated out, the resulting conceptual memory is basically episodic in nature. Definitions of words are part of lexical memory. Consequences of events involving concepts are part of episodic conceptual memory. Associations between concepts are limited to the way

concepts can relate within complete action-based
conceptualizations. Supersets are mostly artificial
constructs with definitions in lexical memory and without a
place in the episodic conceptual memory.

E. Scripts

Some of the episodes which occur in memory serve to
organize and make sense of new inputs. These episodic
sequences we call *scripts*. A script is an elaborate causal
chain which provides world knowledge about an often
experienced situation. Specifically, scripts are associated as
the definitions of certain situational nouns. Words whose
definitions are scripts are, for example, restaurant, football
game, birthday party, classroom, meeting. Some words that
have scriptal definitions have physical senses as well, of
course. "Restaurant", for example, has a physical sense
which is only partially related to its scriptal sense.
The notion of scripts has been proposed generally and
specifically in different forms by Minsky (1975), Abelson
(1973), and Charniak (1972). What we call scripts
represent only a small subset of the concept as used by
others. For our purposes, scripts are predetermined
sequences of actions that define a situation. Scripts have
entering conditions (how you know you are in one), reasons
(why you get into one), and crucial conceptualizations
(without which the script would fall apart and no longer be
that script). In addition, scripts allow, between each causal
pair, the possibility for the lack of realization of that
causation and some newly generated behavior to remedy the
problem. In general, scripts are nonplanful behavior except
when problems occur within a script or when people are
planning to get into a script.
Scripts are recognizable partially by the fact that, after
they have been entered, objects that are part of the script
may be referenced as if they had been mentioned before.
For example:

I. John went into a restaurant. When he looked at the
menu he complained to the waitress about the lack of
choice. Later he told the chef that if he could not make
much, at least he could make it right.

II. We saw the Packers-Rams game yesterday. The Packers won on a dive play from the two with three seconds left. Afterwards they gave the game ball to the fullback.

These paragraphs have in common that they set up a script in the first sentence. This script then sets up a set of roles which are implicitly referenced and a set of props which are implicitly referenced. From that point, roles and props can be referenced as if they had already been mentioned. (Actually, we would claim that they *have* been mentioned by the definition of the script word.)

The script also sets up a causal chain with the particulars left blank. The hearer goes through a process of taking new input conceptualizations previously predicted by the script that has been entered. As before, pieces that are not specifically mentioned are inferred. Now, however, we are going further than we went in the first part of this chapter, in saying that the sequences implicitly referenced by the script are inferred and treated as if they were actually input. This is intended to account for problems in memory recognition tasks that occur with paragraphs such as these.

Let us consider the restaurant script in some more detail.

Script: restaurant
Roles: customer, waitress, chef, cashier
Reason: hunger for customer, money for others

Part 1: Entering
   *PTRANS   (into restaurant)
    MBUILD   (where is table)
    ATTEND   (find table)
    PTRANS   (to table)
    MOVE   (sit down)

Part 2: Ordering
    ATRANS   (receive menu)
    ATTEND   (look at it)
    MBUILD   (decide)
    MTRANS   (tell waitress)
    MTRANS   (waitress tells chef)

DO   (chef prepares food)

Part 3: Eating
   ATRANS   (waitress gets food)
   *ATRANS   (receive food)
   *INGEST   (eat food)

Part 4: Leaving
   MTRANS   (ask for check)
   ATRANS   (leave tip)
   PTRANS   (to cashier)
   *ATRANS   (pay bill)
   PTRANS   (exit)

In the restaurant script given above, we have over-simplified the issue as well as arbitrarily decided the kind of restaurant we are dealing with. Basically, a restaurant is defined by a script that has only the starred ACTs. Restaurants exist where the maitre d' must be tipped in order to get a table, where you get your own food, and so on. The sequence of ACTs above is meant only to suggest the abstract form of a script. A complete script would have at each juncture a set of "what-ifs" which would serve as options for the customer if some sequence did not work out. In addition, other kinds of restaurants could have been accounted for in one script by having choce points in the script. A still further issue is that this is how the restaurant looks from the customer's point of view. Other things happen in restaurants that have nothing to do with restaurants per se and other participants see them differently.

With all these disclaimers aside, the restaurant script predicts all the conceptualizations whose ACTs are listed above, and in a manner quite similar to the functioning of our language analysis program (Riesbeck, 1974) it is necessary in understanding to go out and look for them.

Stories, fortunately for the hearer, unfortunately for us, usually convey information which is out of the ordinary mundane world. The first paragraph tells that the customer did not like the restaurant. His complaining behavior is part of the "what-if" things which we mentioned earlier. The purpose of the script is to answer questions like "Did he eat?" and "Why did he complain to the chef?" These are trivial questions as long as one understands implicitly the script that is being discussed. Otherwise they are rather difficult.

It is perhaps simplest to point out that the second paragraph is incomprehensible to someone who does not know football yet makes perfect sense to someone who does. In fact, football is never mentioned, yet questions like "What kind of ball was given to the fullback?" and "Why was it given?" are simple enough to answer, as long as the script is available.

V. SUMMARY

We are saying that the process of understanding is, in large part, the assigning of new input conceptualizations to causal sequences and the inference of remembered conceptualizations which will allow for complete causal chains. To a large extent, the particular chains which result are tied up in one's personal experience with the world. Information is organized within episodic sequences and these episodic sequences serve to organize understanding. The simplest kind of episodic sequence is the script that organizes information about everyday causal chains that are part of a shared knowledge of the world.

Human understanding, then, is a process by which new information gets treated in terms of the old information already present in memory. We suspect that mundane observations such as this will serve as the impetus for building programs that understand paragraphs.

ACKNOWLEDGMENTS

The research described here was done partially while the author was at the Institute for Semantic and Cognitive Studies, Castagnola, Switzerland, partially while the author was at Bolt Beranek and Newman, Cambridge, Massachusetts, and partially at Yale University.

APPENDIX: CONCEPTUAL DEPENDENCY

We regard language as being a multi-leveled system, and the problem of understanding as being the process of mapping linear strings of words into well formed conceptual structures. A conceptual structure is defined as a network of concepts, where certain classes of concepts can be related to other classes of concepts. The rules by which classes of concepts combine are called the conceptual syntax rules. Since the conceptual level is considered to underlie language it is also considered to be apart from language. Thus the conceptual syntax rules are organizing rules of thought as opposed to rules of a language.

Crucial to all this is the notion of category of a concept. We allow the following categories:

PP--a conceptual nominal, restricted to physical objects only,

PA--a state which together with a value for that state describes a PP,

ACT--something that a PP can do to another PP (or conceptualization for mental ACTs),

LOC--a location in the coordinates of the universe,

T--a time on the time line of eternity, either a point or a segment on the line or relative to some other point or segment on the line,

AA-- a modification of some aspect of an ACT,

VAL--a value for a state.

In conceptual dependency, a conceptualization consists of an actor (an animate PP), an ACT, an object (a PP or another conceptualization in the case of three mental ACTs), a direction or a recipient (two PPs indicating the old and new possessors or two LOCs indicating the old and new directions (often PPs are used to denote LOCs in which case the LOC of that PP is what is meant) and an instrument (defined as a conceptualization itself). A conceptualization can also be an object and a value for an attribute of that object. Conceptualizations can relate in certain causality relations.

We have required of our representation that if two sentences, whether in the same or different languages, are agreed to have the same meaning, they must have identical representations. That requirement, together with the requirement that ACTs can only be things that animate objects can do to physical objects severely restricts what can be an ACT in this representation. We have found that it is possible to build an adequate system (that is, one that functions on a computer for a general class of sentences and that has no obvious deficiencies in hand analysis) using only eleven ACTs.

The eleven ACTs that are used are:

ATRANS: The transfer of an abstract relationship such as possessions, ownership or control. Thus one sense of give is: ATRANS something to oneself. "Buy" is made up of two conceptualizations that cause each other, one an ATRANS of money, the other an ATRANS of the object being bought.

PTRANS: The transfer of the physical location of an object. Thus go is PTRANS oneself to a place, put is PTRANS of an object to a place. Certain words only infer PTRANS. Thus throw will be referred to as PROPEL below, but most things that are PROPELed are also PTRANSed. Deciding whether PTRANS is true for these is the job of the inference program.

PROPEL: The application of a physical force to an object. PROPEL is used whenever any force is applied regardless of whether a movement (PTRANS) took place. In English, push, pull, throw, kick, have PROPEL as part of them. "John pushed the table to the wall" is a PROPEL that causes a PTRANS. "John threw the ball" is a PROPEL that involves an ending of a GRASP ACT at the same time. Often words that do not necessarily mean PROPEL can probably infer PROPEL. Thus break means to DO something that causes a change in physical state of a specific sort (where DO indicates an unknown ACT). Most of the time the ACT that fills in the DO is PROPEL although this is certainly not necessarily the case.

MOVE: The movement of a bodypart of an animal by that animal. MOVE is nearly always the ACT of an instrumental conceptualization for other ACTs. That is, in order to throw, it is necessary to MOVE one's arm.

Likewise MOVE foot is often the instrument of hand. MOVE is less frequently used noninstrumentally, but kiss, raise your hand, scratch are examples.

GRASP: The grasping of an object by an actor. The verbs hold, grab, let go, and throw involve GRASP or the ending of a GRASP.

INGEST: The taking in of an object by an animal to the inside of that animal. Most commonly the semantics for the objects of INGEST (that is, what is usually INGESTed) are food, liquid, and gas. Thus eat, drink, smoke, breathe, are common examples of INGEST.

EXPEL: The expulsion of an object from the body of an animal into the physical world. Whatever is EXPELed is very likely to have been previously INGESTed. Words for excretion and secretion are described by EXPEL. Among them are sweat, spit, and cry.

MTRANS: The transfer of mental information between animals or within an animal. We partition memory into three pieces: The CP (conscious processor where something is thought of), the LTM (where things are stored) and, IM (intermediate memory, where current context is stored). The various sense organs can also serve as the originators of an MTRANS. Thus tell is MTRANS between people, see is MTRANS from eyes to CP, remember is MTRANS from LTM to CO, forget is the inability to do that, learn is the MTRANSing of new information to LTM.

MBUILD: The construction by an animal of new information from old information. Thus decide, conclude, imagine, consider, are common examples of MBUILD.

SPEAK: The actions of producing sounds. Many objects can SPEAK, human ones usually are SPEAKing as an instrument of MTRANSing. The words say, play, music, purr, scream involve SPEAK.

ATTEND: The action of attending or focusing a sense organ towards a stimulus. ATTEND ear is listen, ATTEND eye is see and so on.   ATTEND is nearly always referrred to in English as the instrument of MTRANS.   Thus in conceptual dependency, see is treated as MTRANS to CP from eye by instrument of ATTEND to eye to object.

The states that are used in this paper are *ad hoc*.   A more adequate treatment can be found in Schank (1975).

REFERENCES

Abelson, R. The Structure of belief systems. In R. Schank & Colby (Eds.), *Computer Models of Thought and Language*.   San Francisco: Freeman, 1973.

Anderson, J., & Bower, G., *Human Associative Memory*. Washington, D.C.:   Winston-Wiley, 1973.

Bartlett, F. C. *Remembering: a study in experimental and special psychology*.   Cambridge: Cambridge University Press, 1932, p. 65.

Charniak, E. He will make you take it back: A study in the pragmatics of language (Technical Report).   Castagnola, Switzerland: Instituto per gli studi Semantici e Cognitivi, 1974.

Collins, A. M., & Quillian, M. R. Retrieval time from semantic memory.   *Journal of Verbal Learning and Verbal Behavior*, 1969, 8.

Conrad, C. Cognitive economy in semantic memory.   *Journal of Experimental Psychology*, 1972, 92(2).

Goldman, N., & Riesbeck, C. A conceptually based sentence Paraphraser (Stanford AI Memo 196). Stanford, California: Stanford University, 1973.

Minsky, M. A framework for representing knowledge.   In P. H. Winston (Ed.), *The psychology of computer vision*, New York: McGraw-Hill, 1975.

Quillian, M. R. Semantic memory.   In M. Minsky (Ed.), *Semantic Information Processing*.   Cambridge: Massachusetts Institute of Technology Press, 1968.

Rieger, C. Conceptual memory.   In R. Schank (Ed.), *Conceptual information processing*. Amsterdam: North-Holland, 1975.

Riesbeck, C. Conceptual analysis.   In R. Schank (Ed.), *Conceptual information processing*.   Amsterdam: North-Holland, 1975.

Rumelhart, D. E., Lindsay, P. H., & Norman, D. A. A process model for long-term memory.   In E. Tulving and W. Donaldson (Eds.), *Organization of memory*. New York: Academic Press, 1972.

Schank, R. C. *Conceptual information processing*.   Amsterdam: North-Holland, 1975.

Schank, R. Identification of conceptualizations underlying natural language. In R. Schank and K. M. Colby (Eds.), *Computer Models of Thought and Language.* San Francisco: Freeman, 1973a.

Schank, R. Causality and reasoning (Technical Report 1). Castagnola, Switzerland: Instituto per gli studi Semantici e Cognitivi, 1973b.

Schank, R. Conceptual dependency: A theory of natural language understanding. *Cognitive Psychology*, 1974, *3*(4).

Schank, R. C., Goldman, N. M., Rieger, C., & Riesbeck, C. MARGIE: Memory, analysis, response generation, and inference in English. *Proceedings of the Third International Joint Conference on Artificial Intelligence*, 1973.

Tulving, E. Episodic and semantic memory. In E. Tulving and W. Donaldson (Eds.), *Organization of memory.* New York: Academic Press, 1972.

CONCEPTS FOR REPRESENTING MUNDANE REALITY IN PLANS

Robert P. Abelson
Yale University
New Haven, Connecticut

I. INTRODUCTION

As a social psychologist, I am heartened by the direction taken by the work appearing in this book. Machine analysis of language has become more "psychological", in that many artificial intelligence workers are asking how a program can make inferences with the scope and flexibility (and risk of being wrong) characteristic of human language understanding. This emphasis is congenial to me because I

273

have long been struggling with the problem of how to represent structures of belief about social issues. In previous theoretical papers (Abelson, 1968, 1973; Abelson & Reich, 1969) I have pointed to the importance of higher-level implicational constructs.

Ideological systems are an especially interesting type of knowledge system in that they are extremely "hypothesis-driven" rather than "data-driven" (D. Bobrow & Norman, Chapter 5). The ideolog is apt to find evidence even in the most commonplace events for the predicted workings of his enemies, whether pinks or pigs. In the Cold War ideological structure (analyzed in detail by Abelson, 1973) an extraordinary number of actual and potential international events are "understood" as confirming the long range objective of the Communist bloc to control the world, and the need for Free World counteraction. In like structural manner, radical revolutionary ideology focuses on the machinations of an oppressive imperialism which must be overturned. Belief systems which are less than totally ideologized, while they may not use the same single core account for everything, nevertheless often have the "top-down" character of imposing abstract and possibly gratuitous interpretations on the ambiguous data presented by the world of events.

It is intriguing to study how such heavily hypothesis-driven systems work, and among the many methods of study is computer imitation of possible mechanisms. When we tried to create a computer program which would simulate an ideolog (Abelson & Reich, 1969), however, we soon discovered that the system could not interpret events sensibly without some reasonable level of mundane knowledge about simple physical properties of persons and objects, independent of ideological import. Thus our system, operating entirely at an abstract level, reasoned that since Latin American radical students had thrown eggs at Nixon (as Vice President, visiting in 1958), it was quite plausible that Fidel Castro might throw eggs at Taiwan. The system anticipated continual hostile provocation by Communist actors against vulnerable Free World targets, since that was a key presumption in its master belief. It was, however, innocent of the logistics of egg throwing and of the relative vulnerability of persons versus islands as egg targets (as well as many other low-level facts), so that it often emitted output statements foolishly out of touch with physical reality. Being fanatically ideological is quite different from being schizophrenically disoriented. A convincing simulation of the cognitive system of an ideolog must combine a solid concrete knowledge base with its highly subjective interpretations of the deeper motives and meanings lying behind simple events.

A. Scripts, Themes, and Dremes

The present chapter revises and fills in my earlier theoretical system (Abelson, 1973) linking simple actions and higher-level constructs. The terminology[1] I will use is as follows: a *script* is a conceptual structure which explains for the believer why a specific social action or sequence of actions has occurred or might occur; a *theme* is a conceptual structure which accounts for a number of related scripts involving the same actor or actors--a theme is thus a script-generator, based on perceived enduring characteristics of actors or relationships between actors; a *dreme* is a conception of the possibility that one or more themes are subject to change--that a system of enduring characteristics or relationships need not endure forever.

To illustrate with the Cold War ideology, a given script might concern Communist military action against South Vietnam; the explanatory theme for this could be the Communist project of taking over all of Southeast Asia by eroding Free World resistance; the relevant dreme would be for the Communists to abandon this project some day (because of Free World determination and power), nevermore to participate in themes of conquest.

---

[1]In the earlier chapter, I used the term *script* at a higher level than it will be used here. The demotion of this term seems more in keeping with the common understanding of a script as a sequence of particular events.

B. Discussion of General Theoretical Strategy

Work with language-understanding knowledge or belief systems always runs into the "size problem" in one way or another. There is much too much common sense knowledge of the world in even the humblest normal human head for present computer systems to begin to cope with. One can adjust to this problem by deliberately restricting the computer system to deal with a very small artificial world, as Winograd (1972) did with his famous "toy world" of blocks on a table top. Another strategy is to design very general mechanisms so that in principle the computer might be able to store and make intelligent inferences from almost anything you tell it. This was the guiding orientation of Rumelhart, Lindsay, & Norman's (1972) early memory model, and the same type of strategy is evident in Rieger's (1974) very rich machinery for conceptual inference. If one goes this route, however, the memory cannot be loaded with too much variety of detail about too many things before it exceeds reasonable storage bounds. Still, one can study the mechanisms in principle and not worry too much about operating practicalities.

Many present views, including my own, lie somewhere in the middle of the "small world, complete knowledge" versus "big world, scattered knowledge" dimension. One place to find a stylized world with moderate size, moderate realism, and substantial interest is in the specialized lexicon and stereotyped social action patterns of an ideology. Another is in stylized literary forms such as fairy tales, myths, Westerns, "true romances", etc. If, however one uses natural language input, then all the complexities of (say) English remain to bloat the memory. One way to forestall this is to operate at a level of abstraction above individual words. Schank (1972, 1973) has convincingly shown with his conceptual dependency analysis that is is possible to separate concept representation from the problems of words and word senses, and to deal productively with the interface between them (Schank, Goldman, Rieger, & Riesbeck, 1973; Schank, 1975; Goldman, 1974, Rieger, 1974; Riesbeck, 1974). In preparing the present chapter, I have profited greatly from several extended discussions with Schank. His approach and mine are nicely complementary, and a unified system including both seems quite feasible.

In this chapter I will concentrate on *scripts*, because theory requires a good foundation at the script level. The simplest relevant conceptual *script* content includes the purposeful transactional activities of social actors such as communication with others, enlisting the help of others, and so on; and a set of specifications of the realities constraining the possibilities of action--knowledge of transportation and communication systems, the necessary properties of objects, etc.

Of special theoretical concern are conditions which prevent intended acts or cause them to fail. Although we will not discuss *theme* concepts in detail in this chapter, a central thematic idea is that actors seek to frustrate the purposes of their adversaries. Often this takes the form of selective physical disruption of crucial activities (e.g., by blockade, sabotage, boycott, etc.); thus we will examine activities carefully for their points of vulnerability, particularly in the instrumentalities by which actions are carried out. This emphasis is slightly different from other artificial knowledge systems, which do not tend to concentrate on the many ways simple actions can be disrupted. Nor do they tend to be much concerned with social acts such as persuasion, which I will broach in a very rudimentary way.

Two final caveats are in order before plunging into low-level details. Our *scripts* are "pure" conceptual structures representing abstract event chains. They do not mirror the actual flow of reality, with its kaleidoscope of intersecting happenings; rather, they reflect how a cognizer might imagine a slice of reality would proceed, barring interruptions and irregularities. They are structures in the mind of the script's protagonist, or more precisely for our purposes, structures *imputed by an observer* to his mind. In Schank's terms, they are the inferred details of MBUILDs. Very often, an observer understands human actions by seeing them as part of some plan or maneuver the observer thinks the actor has in mind. Our belief system model thus views the world like an observer of purposive operations, rather than like an actor itself.

Neither do our *scripts* correspond to stories or fables, such as those analyzed in the chapters by Rumelhart (Chapter 8) and Schank (Chapter 9). The scripts of the protagonist(s) in stories are interrupted by properties of the situation which induce new scripts and themes. There are deep theoretical issues to be worked out in coordinating "planning scripts" and "situational scripts". A start in this direction has been made by Schank & Abelson (1975).

II. PRIMITIVE STEPS IN PLANS

The major type of *script* is a *plan*, specified by the actor, the goal state, and the sequence of intended steps to reach the goal. In this chapter, we promote the point of view that the steps-in-plans are based on a small set of primitives and occur in highly constrained ruleful sequences.

Schank's (1973) work makes clear that with the aid of conceptual dependency formalism a small set of primitive act concepts (MOVE, GRASP, PROPEL, INGEST, EXPEL, SPEAK, ATTEND, MBUILD, MTRANS, PTRANS, ATRANS) is sufficient to represent almost any describable event (see the appendix in Chapter 9 for a description of these ACTS). In a previous work (Abelson, 1973) and in more recent unpublished work, I attempted to use a bastardized Schankian system of acts for describing the steps in plans. What I did was to ignore the mechanical physical acts (MOVE, GRASP, etc.) as being too low level, but incorporate variants of the more abstract transactions (MTRANS, PTRANS, ATRANS) and add a couple of new asbstract transactions, e.g., "QTRANS", a change in the quality of some object. I have not heretofore analyzed why bastardization of Schank's primitive act set seemed necessary, but a clarification of that issue is now in order.

Acts as steps-in-plans have two special features distinguishing them from acts as events-in-reality. Both features stem from the intentionality of plans. First, the reason why low-level primitives such as GRASP are not the most useful components in the cognition of plans is that for human actors they are usually not problematic--they are taken for granted by both planner and observer. They are "basic actions" or "gifts" (Danto, 1965; Boden, 1973) which typically cannot be introspectively analyzed as to how they are done. It is therefore not sensible to speak of intending a basic action such as grasping unless some abnormality has rendered it nonbasic, or the speaker wishes to use irony to emphasize his natural intent. Thus, "John planned to lift his arm" implies disability or danger in so attempting, while "I am going to grasp this door knob" is either an empty statement or a determined announcement of an intent to go out (or in). We leave these nuances to linguists and philosophers, and systematically underplay basic actions as plan steps. Of course, that is not to say that the execution of plans does not require grasping, ungrasping, moving, etc. Indeed, this kind of nitty-gritty is mostly what robot programs such as Winograd's (1972) SHRDLU do. Nevertheless, we are distinguishing the *analysis* of plans from the execution of plans, and for this we need the steps consciously intended by the actor. If the actor is a robot in a limited world, then the analysis may indeed be at a low level. If the actor is a human actor in the everyday world, however, higher-level steps are needed, with lower-level executions packaged as subroutines (unconscious, so to speak) within higher-level plan programs.

The second consideration is closely related to the first. Acts as steps-in-plans are characterized by their intended effects rather than by their physical nature. Consider the earlier example of Latin American radicals throwing eggs at Nixon. From a physical standpoint, the egg-throwing is a propelling action performed by the radicals on the eggs in the direction of Nixon. From an intentional standpoint, the egg-throwing is instead classifiable according to its design: namely, to mess up Nixon's neat appearance. Such an intentional operation might be seen as a member of the class QTRANS, changing the quality of an object (in this case, changing Nixon from neat to messy-looking). To be sure, the propelling action is a *part* of this QTRANS-- specifically, it is the instrumentality for accomplishing the intention. The major focus, however, belongs at the QTRANS level, because other instrumentalities could be substituted for egg-throwing in achieving the same purpose.

An explication of "intended effects" sharpens the analysis. Since every "effect" can be viewed as some state of affairs which did not obtain prior to the operation in question (otherwise no operation would have been necessary), we may say that *every step in a plan corresponds*

TABLE I.
States and Actions Which Change Them

LABEL Explanation	Description

1 (a) PROX(X,Y)   State of *proximity*
Entity X is proximate to entity Y.
  (b) ΔPROX(A,X,Z,Y,M)   Change of *proximity*
Actor A causes entity X (possibly himself) to change from proximity to Z to proximity to Y, with means M.

2 (a) HAVE(X,Y)   *Having* something
The possessor of object X is Y.
  (b) ΔHAVE(A,X,Z,Y,M)   Change of *having*
Actor A causes object X to change possessors, from Z to Y, by means M.

3 (a) KNOW(X,Y)   *Knowing* something
Conception X is known by person Y.
  (b) ΔKNOW(A,X,Z,Y,M)   Change of *knowing*
Actor A causes person Y (possibly himself) to know conception X from source Z, with means M.

4 (a) QUAL(X,Y)   Having a *quality*
Object X has quality Y.
  (b) ΔQUAL(A,X,Z,Y,M)   Change in a *quality*
Actor A causes object X to have quality Y instead of quality Z, using means M.

5 (a) OKFOR(X,M)   *Being 'OK'* for an activity
Entity X is in good condition for means M.
  (b) ΔOKFOR(A,X,L,M,N)   Getting to *be OK*
Actor A causes entity X to be in good condition for means M (rather than means L), using means N to do so.

---

to a desired state change. Steps will therefore be labeled according to the state they change, notated by the symbol "Δ", "delta", preceding a state name. To distinguish this level of action concepts from physical acts, we will call them *deltacts*.

Analysis of plans using state concepts and acts which change states is not an original proposal (see Simon, 1972).

TABLE I (cont.).
States and Actions Which Change Them

LABEL Explanation	Description

6 (a) AGENCY(X,(D,Y))   State of *agency*
Person X is an "agent" of person Y for act(s) D.
(b) ΔAGENCY(A,X,(T,Z),(D,Y),M)   Change of *agency*
Actor A causes person X to become an agent of person Y
(usually A himself) for act(s) D, instead of alternative act(s) T
as agent of Z, using means M.

7 (a) LINK((U,C),(Z,Y))   State of *linkage*
There is a "link" through medium U for carrier or channel C,
over the range from Z to Y.
(b) ΔLINK(A,(U,C),(Z,W),(Z,Y),M) Change of *linkage*
Actor A causes the range of medium U for carrier C to change
from (Z to W) to (Z to Y), using means M.

8 (a) POWER((A,M),(S,E))   Availability of *power*
Actor A has available for means M the amount E of power
from source S.
(b) ΔPOWER(A,M,(S,E_1),(S,E_2),N) Change amount of *power*
Actor A causes a change in the availability of power for means
M from source S from E_1 to E_2, using means N.

9 (a) UNIT(X,Y)   Existence of a *unit*
Object X occurs in a unit with object Y.
(b) ΔUNIT(A,X,(X,Z),(X,Y),M)   Change in the *unit*
Actor A causes object X to enter unit connection with object Y
rather than object Z using means M.

The trick lies in what set of states and deltacts is
postulated for social actions. A particular set is here
suggested. The crucial feature for which we have striven is
that the set be general over many contexts, yet still small
and "closed". The deltacts are all defined as operations
which cause state changes. For the system to be fully
closed, it is also necessary that the conditions needed to
enable performance of the deltacts arise naturally from the

same set of states. A plan consists in the selection of one
or more deltacts which change states so that it is possible
to perform other deltacts which change states so that,...etc.,
until the final goal state is reached. Table I summarizes
the proposed set of interrelated concepts. After a general
introduction of these concepts, we will discuss their
features individually, and then show how they string
together in the enablement and causation chains
constituting a plan. In an earlier draft of this chapter, an
attempt was made to present our ideas in highly formal
notation, but the sacrifice in readability proved to be too
great. Therefore, in the present version we will concentrate
on verbal explication of the concepts. A technical account
of how a LISP program using our rules can generate
meaningful plans will be presented in a separate technical
publication.

The first three deltacts listed in Table I are closely
related to three of Schank's acts: PTRANS, MTRANS, and
ATRANS. For each there are five arguments: the actor,
the object acted on, then two arguments denoting the from-
to conditions (directional or recipient cases, in Schank's
terms)--we call these the *origin* and *terminus*, respectively--
and finally, the instrumentality or means by which the
operation is carried out. The corresponding states have two
arguments, one for the object, and one for the "value" the
object assumes at the terminus of the deltact. For example,
in a ΔPROX, the actor A causes object X, originally
proximate to Z (either a fixed location or another object)
to become proximate to Y (either a fixed location or
another object), e.g., by carrying the object, throwing it,
shipping it, etc. Thus the original state PROX(X,Z) is
changed to PROX(X,Y).

The next several states depart from Schank's. The
concept discussed above as QTRANS is here called ΔQUAL
to conform to the delta notation. Many different types of
object qualities could be included in the lexicon of the
knowledge base, and the blurring of distinctions among
them with our global deltact would have to be corrected by
detailed discriminations at the *means* level.

One particular "quality" of importance in all sorts of
enabling conditions is the catch-all state OKFOR. We
designate an object as OKFOR an intended means of a

deltact if it is in all respects in appropriate condition for that means; and an actor as OKFOR a given means if his health and state of mind are adequate to the task. The class ΔOKFOR thus includes a variety of fixing and restorative operations, which can be viewed as special cases of ΔQUAL.

The next state, AGENCY, is very crucial in our system. It is a state of induced intention for an actor to perform one or more acts under the aegis of a second actor who wants those acts performed. There are thus three arguments for AGENCY: the two actors and the induced act. In order to induce AGENCY, the second actor must persuade or coerce the first actor. Actions in this broad class constitute ΔAGENCY.

Rewards and punishments are often used in the service of ΔAGENCY. These constitute forces evoking the induced actions. Since these forces may need to be sustained over time, it is perhaps more appropriate to speak of the POWER sustaining an induced action. We represent this with a POWER state, using as arguments an actor-act unit, and a source-amount unit, specifying how much of a particular type of motive power has been used. In addition to psychological POWER, physical POWER is sometimes important, as explained later. By ΔPOWER we designate those actions which increase (or decrease) the amount of POWER.

The next state listed in Table I is what we call LINK. It signifies a preexisting connection between two locations through a "medium" (air, road, wires) so as to permit a transportation or communication operation, i.e., ΔPROX or ΔKNOW, using a "carrier" system (airline, car, phone system). The set of existing LINKS in a given geographic domain effectively constitutes a functional map of that domain. As mentioned previously, it is important for a belief system to have this kind of low-level knowledge in order to separate the physically possible from the physically impossible. Generally, LINKS are relatively enduring, but we include the deltact ΔLINK in our system to handle a variety of cases of implementation or interruption of LINKs.

The last state, UNIT, expresses the idea of the attachment (containment) of a physical object to (in) another, or the joining of forces of two or more actors, such that the conjunctions of objects or actors share common fate (see Heider, 1975). Three men on a horse constitute a UNIT, as do a letter and its mailing envelope. The operation ΔUNIT represents the assembling together and appropriate joining of the objects to be formed in a UNIT, or else the disjoining of objects previously in a UNIT.

This system of nine states--PROX, KNOW, HAVE, QUAL, OKFOR, AGENCY, POWER, LINK, and UNIT-- together with their corresponding deltacts are designed as a closed set for expressing scripts of steps-in-plans. Our scheme will make more sense to the reader as we take up each state and deltact in more detail, in conjunction with the concept of the *means* by which deltacts are performed, and the formal relations CAUSE, ENABLE, SUCCEED, and FAIL. The system is evolving, and no doubt will change as new problems and opportunities are revealed. In later sections we will discuss some conceptual problems bypassed (but not forgotten) in this chapter.

III. SPECIFIC STATES AND DELTACTS

A. Proximity

1a. PROX: The state of proximity of one entity to another is conceptually important because it is so often a necessary condition for the enablement of action. Proximity has several obvious formal properties. It is a relation which is apparently reflexive, symmetric, and transitive. These formal properties are, however, not terribly useful in and of themselves, and may lead to misleading conclusions. Mr. Cohen being proximate to Chicago means something different from Mr. Cohen being proximate to his Aunt Minnie's house in Chicago. The "derivation"--

PROX(COHEN, CHICAGO) and PROX(MINNIE'S HOUSE, CHICAGO)

implies PROX(COHEN, MINNIE'S HOUSE) --is not literally appropriate, even though in many contexts we may assume that Cohen can find his way to Minnie's house when in Chicago. One way to cope with this is to associate with every entity an "effective proximity radius", that is, a rough index of how narrow is the range of spatial locations for entering into usual interactions with the entity. When entities with different radii are proximate, the one with the smaller radius determines the transitivity properties of the pair. (This rules out the inappropriate CHICAGO example above). In any case, proximity would ordinarily not be inferred, it would be *produced* by a ΔPROX.

1b. ΔPROX: It is useful to distinguish movable from immovable entities. Mohammed and the mountain notwithstanding, fixed locations or LOCS are not subject to ΔPROX. Movable entities (people and things) can be brought PROX to LOCS by ΔPROX, or PROX to one another by both being ΔPROXed to the same LOC.

Recall that ΔPROX has as its five arguments the actor A, the object X, the origin Z, the terminus Y, and the means M. The possible means include unaided methods for going from one place to another, e.g., walking, but the more interesting cases are those involving some system of transportation. Different limited worlds have different typical means of transportation, ranging from horses to broomsticks to airlines. ('The particular knowledge associated with a particular environment such as the Wild West is what Charles Rieger[2] calls "coloring".)

There are problems in representing the complexities of nonindividual travel, such as by commercial airline. The airline behaves in a way which is (more or less) predictable by the traveler, but he is not the main actor in the means for ΔPROX--that is, he does not fly the plane. He takes advantage of the operating properties of the system, and the system in turn obliges the traveler, provided they get paid. We represent all this within the framework of AGENCY. The airline is in potential AGENCY to the traveler for a particular trip, actualized if he supplies the POWER of an amount of money appropriate to that trip.

---

[2]Personal communication

Whether or not a trip between Z and Y exists at all via a given airline C depends on the condition LINK((U,C),(Z,Y)), interpreted as whether the airline schedules the route (Z,Y).

Even below the complexity level of commercial carriers, the cooperative transfers of initiative from one actor to another for the purpose of actions such as ΔPROX are a genuinely problematic aspect for artificial knowledge bases. Thus, in operations such as General Washington's crossing of the Delaware or a Soviet spy heading for the U.S. by submarine, it is clear to a human interpreter that General Washington does not himself row the boats nor the spy himself operate the submarine. There are predictable institutional ways by which the logistics are accomplished. These ways should be represented with known detail in an artificial system rather than simply be assumed as normal unspecified functions; it may be important to understand the consequences (and the antecedents) of failure of transport systems, as when the submarine's navigator is a counterspy or General Washington's boat has a hole in the bottom.

The deltact ΔPROX, like other deltacts, requires certain enabling states. The following is a somewhat simplified list of conditions which enable ΔPROX(A,X,Z,Y,M):

(a) LINK((U,C),(Z,Y))--there is a link from origin to terminus via the carrier system C through medium U.
(b) AGENCY(C,(M,A))--the carrier is willing to perform means M for the actor. (Actually, this condition is too simple; the carrier, if a large system like an airline, will in turn designate an agent--the pilot(s)--to perform the means.)
(c) PROX(A,Z)--the actor must be at the starting point.
(d) PROX(X,Z)--so must the object to be moved (which might be the actor himself).
(e) HAVE(I,C)--the carrier system must have an instrumental device I (e.g., a particular plane) to perform means M.
(f) UNIT(X,I)--the object must be connected to the instrumental device (e.g., the person must be inside the plane).
(g) OKFOR(X,M)--the object must be in appropriate

condition for the means (e.g., not too big or too heavy so that condition (f) is impossible.)

(h) OKFOR(I,M)--the instrumental device must be in good condition (e.g., flightworthy).

(i) OKFOR(C,M)--the carrier is in normal condition for the means (e.g., not on strike or closed by bad weather); if the carrier is simply the actor himself, say, going by foot, then this condition indicates his ability to make the voyage.

These last three conditions emphasize the range of uses of OKFOR. In any given case, of course, the computation of OKness depends on knowledge of what qualities an object or system has, and knowledge or computation of the significance of those qualities for the means at hand. "If a rowboat is red, is it OK for boating?" "If a horse is three-legged, is it OK for riding?" A sophisticated action knowledge system would need to be able to answer such questions. For our present purposes, we are willing to accept crude algorithms for such computations (e.g., IF QUAL(*object*, DAMAGED), THEN NOT-OKFOR (*object*, *means*.any); etc.).

The next formal relations to consider are the effects of the ΔPROX. *We do not automatically assume that a deltact is successfully completed even though it is apparently adequately planned*; failures could occur in transit. The format of this representation in general is:

IF ⟨success conditions...⟩THEN CAUSE(*deltact*,⟨effects...⟩),
ELSE FAIL(*deltact*)

In the case of ΔPROX, a simplified list of *success conditions* might be:

(a) $E \geq E_0(S,M,Z,Y)$, where POWER((C,M),(S,E))--that is, the power available to the carrier for the means is of sufficient quantity to exceed the threshold required to go from Z to Y by that means.

(b) OKFOR(S,M)--the power source is in appropriate condition (e.g., there are no ice pellets in the gasoline).

It is important to realize that plans can vary according

to the number of enabling conditions considered by the planner. If a planner cuts corners by presuming enablements true when they might not be, this has the effect of moving enabling conditions to the success conditions list.

If these conditions (and the enablements) are met, then ΔPROX *causes* (at least) the following effects:

(a) PROX(X,Y),
(b) NOT-PROX(X,Z).

If these conditions are not met, ΔPROX fails. Such failure would of course have important side consequences (e.g., the plane crashing or landing in a corn field).

We think it important to make the distinction between enablement conditions and success conditions, though both are necessary for a deltact to produce its intended effects. Enablement conditions are necessary in order that the operation begin at all (if you do not have a ticket, you cannot board the plane), and success conditions for the act to be completed properly once begun (if there is not enough fuel to get to Chicago, the plane cannot land there). This distinction is sometimes arbitrary, as there are circumstances under which a properly supervised operation *should not* begin (e.g., not having enough gasoline, or the pilot being drunk) yet might nevertheless be initiated. The distinction could be very important to a potential adversary wishing to intervene to thwart the plan. If the adversary wishes merely to hold up the plan, and anticipates a coming operation, he may remove an enablement of that operation. If, however, his motives are more violent and he wishes to bring the plan to disaster, or his timing is such that he has missed the opportunity for prevention, he might try to bring about a failure of an operation in progress.

B. Having

2a. HAVE: The state of having is often an enabling condition in one way or another for deltacts. In the context of this usage, what we intend by HAVE here is often more transient than ownership; it is possession for instrumental use, or else mere passive possession. For convenience, when an object X is (temporarily) possessed by no one, we say HAVE(X,"public domain"). The public domain is like a giant banker yielding things to and receiving things from individual actors.

2b. *ΔHAVE:* In ΔHAVE, actor A causes object X to change possessors from Z to Y. Two cases have been compacted in the same notation: a "giving" or "losing" action in which main actor and original possessor Z are identical (while final possessor Y is different); and a "taking" or "finding" action in which the main actor and final possessor Y are identical (while original possessor Z is different). We note in passing that although ΔHAVE is similar to Schank's ATRANS, finding from the public domain is not an ATRANS because there is no human donor. Instead it is a PTRANS. There are several small differences of this kind between the "Δs" and the "TRANSs", in addition to the major difference that the Δs describe projected operations in the mind of the planner, rather than actions in the physical world.

The means of ΔHAVE can be of two basic forms: one by which the object is essentially transferred hand-to-hand, and the other by which the object is sent to possessor Y from possessor Z (=*actor*). This sending is some form of ΔPROX, often by means of an institutional mailing system. In such cases where a means is itself a deltact, then the *enable* and *success* conditions for this "inner" deltact must also hold in order for the "outer" deltact to succeed.

The simplest *enable* conditions for ΔHAVE(A,X,Z,Y,M) itself are merely:

(a) HAVE(X,Z)--by definition, the original possessor has the object.
(b) PROX(A,Z)--the actor is proximate to the original possessor (who might trivially be himself)

This makes it all look easy, but in the case of the means of sending, the nine enabling conditions for ΔPROX are also involved.

The *success* condition for ΔHAVE itself is simply (a) below, with (b) (and three ΔPROX conditions) also needed for the means of sending the object from some location K to some location L.

(a) OKFOR(Y,M)--the receiver has to be OK for the transfer means, which might mean being nimble or strong if theft is involved, but simply being alive if he is just a passive recipient.
(b) PROX(Y,L), where M=ΔPROX(A,X,K,L,N)--that is, the receiver must be at the place to which the object X is sent. (The enabling conditions for the ΔPROX include the complementary condition PROX(A,K)).

If these conditions are met, then ΔHAVE *causes* (at least) the following effects:

(a) HAVE(X,Y),
(b) NOT-HAVE(X,Z).

C. Knowing

3a. KNOW: The state of knowing is subject to many subtleties, in particular the questions of whether KNOW differs from BELIEVE, and whether KNOW can take KNOW as an argument, with KNOW as an argument, etc., like barber shop mirrors. We bypass the first of these subtleties, treating KNOW and BELIEVE as the same, but accept any concept as second argument of KNOW.

3b. ΔKNOW: The specifications for ΔKNOW are familiar because most of the lower-level details are isomorphic with those of ΔPROX, except that a communication channel rather than transportation carrier is involved. Something is changing "location", but it is "mental location" (Schank, 1972) rather than physical, and the something is a *conception* rather than an *object*. Analogously with ΔHAVE, the actor corresponds to either the source or the recipient.
It is a *coded form of the conception* (such as a telegram) rather than the conception itself which travels through the medium of communication. A detailed knowledge system would presumably store the typical forms of information codes attaching to various communication systems. Another

special issue arising with ΔKNOW is that the origin of the information transfer could be physical storage (like a book or a tape recording) rather than the mind of an individual. In the case where the origin is physical storage, the actor usually corresponds to the recipient, and the means consists of reading the information out of the source with the possible aid of a decoding instrument.

The following conditions *enable* ΔKNOW(A,X,Z,Y,M):

(a) LINK((U,C),(Z,Y))--there is a link from source to terminus via communication carrier system C, through medium U from Z to Y.

(b) AGENCY(C,(M,A))--the carrier system is willing to perform means M for the actor.

(c) PROX(A,C)--the actor is at a terminal of the carrier (e.g., a phone booth).

(d) KNOW(X,Z)--the source Z must originally know the conception X; alternatively, if the source is inanimate information storage, then this condition is UNIT(F,Z), where F is the physical form of X.

(e) HAVE(I,C)--the carrier must have the necessary instrument to perform means M.

(f) OKFOR(F,M)--the form of the information is appropriate for transmission by means M.

(g) OKFOR(I,M)--the instrumentation is in good condition.

(h) OKFOR(C,M)--the carrier system is in normal order.

The success conditions for ΔKNOW correspond to the conditions (a)-(c) for ΔPROX, plus the further condition:

(d) OKFOR(Y,M)--the recipient must be capable of receiving the information by the indicated means (e.g., not be deaf if telephoning is the means).

If all conditions are met, ΔKNOW causes KNOW(X,Y).

D. Quality

1a. QUAL: As remarked previously, QUAL(X,Y) is a catch-all state for many conceivable qualities of objects. The only useful general inference principle for QUAL obtains by using the formal relation of set membership, ISA, extensively employed among others by Rumelhart, Lindsay, & Norman (1972) in discussions of semantic memory. If QUAL(Z,Y) and ISA(X,Z), then QUAL(X,Y). More important for our purposes, however, are the dynamics of how new qualities are produced.

4b. ΔQUAL: For a ΔQUAL, the critical features depend on the means. To change the quality of a dragon from ALIVE to DEAD, you run a sword through him; to change a princess from NORMAL to ENCHANTED, you feed her a poisoned apple. Most such means involve at least one instrumental device (sword, apple, etc.). Additionally, a certain quantity of power from a particular source may be required: to change meat from RAW to COOKED, a certain amount of heat from a fire (or oven, etc.) is necessary. These concepts enter the conditions lists.

The *enable* conditions for ΔQUAL(A,X,Z,Y,M) are:

(a) QUAL(X,Z)--the object X has initial quality Z.
(b) PROX(A,X)--the actor must be near the object to be altered.
(c) OKFOR(A,M)--the actor must be in fit condition for the means (e.g., the knight robust enough for dragon killing).
(d) OKFOR(X,M)--the object must be in condition permitting the means.
(e) HAVE(I,A)--the actor must have the instrumental device, if any, for performing the means.
(f) OKFOR(I,M)--the device must be in good condition.

The *success* conditions are:

(a) $E \geq E_0(S,M,Z,Y)$, where POWER((A,M),(S,E))--the power applied in the means must be sufficient to change the quality of X from Z to Y.
(b) OKFOR(S,M)--the power source is in good condition.

If the conditions are met, ΔQUAL causes QUAL(X,Y) and NOT-QUAL(X,Z).

E. Being OK For

5a. OKFOR: Formally, OKFOR(X,M) is a conjunctive condition; ordinarily, any single relevant thing wrong with X will spoil it for use. One way to represent this is to write: OKFOR(Q(X),M) for all qualities such that QUAL(X,Q), implies OKFOR(X,M).

5b. ΔOKFOR: The preceding statement suggests, then, that the way to make an object OKFOR a means is to improve its inadequate qualities. This involves the deltact ΔQUAL according to the rule: ΔOKFOR(A,X,L,M,N) = ΔQUAL(A,X,Z,Y,N) for all (Z,Y) such that NOT-OKFOR(Z,M) and OKFOR(Y,M). (Here the argument N represents the means-set used for the ΔQUALS, and L is an argument representing what X was OKFOR--probably nothing--before the change to being OKFOR means M).

Thus we do not need new conditions for ΔOKFOR, since it is a special form of ΔQUAL. One other feature of ΔOKFOR deserves mention, however. Like all the other deltacts, only more obviously so, the Δ is directional with respect to the achievement of a plan--that is, we are speaking of change in the service of a plan. We could also have *negative* Δs, that is, deltacts which thwart plans, in this case Δs which render objects NOT-OKFOR means. This issue is discussed in a later section.

F. Agency

6a. AGENCY: The purpose of having a state of AGENCY in our system is that plans often involve subcontracting of certain actions to other individuals. In subcontracting, the linearity of intention is preserved, even though the actor changes. This condition is important, because a change of actor (from one sentence to the next) usually indicates a second center of autonomy--the second actor is pursuing his own plan, or at least is an independent source of (re)action. If, however, there has been a specific prior social contract between the two actors, then we are able to understand that the second actor may be acting on behalf of the first. An interesting property of "acting on behalf of" is that it can be transitive. AGENCY(Z,(D,Y)) and AGENCY(X,(D,Z)) imply AGENCY(X,(D,Y)); that is, if Z is primed to do D for Y, and X is then primed by Z to do D (because Z is unwilling or unable to do it himself), then X is in effect primed to do D for Y (though neither need be aware of this).

6b. ΔAGENCY: In bringing about a change in agency, actor Y convinces or induces the target actor X to try to carry out act D in place of whatever he might have been going to do otherwise--act T (which in many cases might be a null category). Actor Y uses a conjunction of two means: he provides some *quantity* of *inducement* as either an incentive or a threat and also informs actor X that he wants the act performed. The informing action is a particular ΔKNOW, carrying with it its own enablement and success conditions (which we will not list separately).

We concentrate here on the inducement means. Of course a friendly prior relationship between the two actors may make it likely that actor X will freely do actor Y the favor D, or actor X may feel he owes actor Y the favor in exchange for something actor Y has done for him in the past. We will not deal with these thematic considerations here. Apart from the history of relationship between the two actors, however, the likelihood that actor X will agree to attempt D depends mainly on the sufficiency of the size of the inducement (its psychological "power"), relative to what is being asked. Whatever influence Henry Kissinger has had from time to time in negotiating with various Arab countries, he did not achieve it by offering them cases of Mogen David wine.

The *enable* conditions for ΔAGENCY (A,X,(T,Z),(D,Y),M) (beyond those in the ΔKNOW by actor Y) are these:

(a) AGENCY(X,(T,Z))--the target actor has some initial intended acts on behalf of actor Z (possibly himself); this is an empty definitional condition except insofar as the nature of the alternative T enters into Xs disposition to be influenced by Y (see below under "success conditions").

(b) OKFOR(A,M)--the initiating actor is in a condition which permits the use of inducement M.

(c) HAVE((S,E),A)--the inducer has at his disposal the amount E of inducement from source S.

(d) OKFOR((S,E),M)--the inducement is in good condition for the act of persuasion (e.g., is not counterfeit, malfunctioning, etc.).

The *success* condition is:

(a) $E \geq E_0(S,T,D)$ where POWER((A,M),(S,E))--that is, the incentive ("power") available to the persuader is of sufficient quantity to exceed the threshold required to convince someone to give up T and do D. This threshold might also depend on X the particular agent, Y the particular client, and Z the previous client. Many thematic complexities may thus enter this threshold condition.

If all conditions are met, then ΔAGENCY causes:

(a) AGENCY(X,(D,Y)),
(b) NOT-AGENCY(X,(T,Z)).

G. Linking

7a. LINK: The LINK state is the one which permits "map" information to be represented qualitatively. The LINK relation is reflexive, (usually) symmetric, and transitive if the carrier and medium are held constant. For example, the endpoints might be cities, the medium highways, and the carrier personal automobiles. Transitivity guarantees the inference that the endpoints of a chain of cities successively linked by highway will also be linked by highway. Such extrapolation of local linkages seems roughly to correspond to the ways people think about trips ("We will drive to Winnipeg, and then we will stop in Regina, and then . . ."), and thus has a more psychological flavor than would a literal pictorial representation of a map. Of course many trips involve the concatenation of different carriers (limousine to the airport, then fly, then take a taxi . . .). A transitivity formalism could readily be established in the mixed carrier case, too, although one would want to be careful about the "radius" problem (New York to Chicago to Minnie's house) discussed previously for the PROX relation. Again, the most specific element in a chain forces its level of specificity on the other elements.[3]

   7b. ΔLINK: There are two subclasses of ΔLINK, one in which LINKS are formed (by building a bridge, closing a phone circuit, etc.) and the other in which LINKS are broken (by destroying a bridge, blocking a road, cutting phone lines, etc.). The two cases are obviously quite different in instrumentality. The way we create abstract generality is to define the origin of the ΔLINK as a certain range (Z,W) between two endpoints of a medium, and the terminus as a different range (Z,Y), stretching between one of the original endpoints and one new endpoint. If the endpoints are locations, the change of range could denote either lengthening or shortening. Thus the Ho Chi Minh trail, for example, was open from North Vietnam either all the way to South Vietnam, or only part way into Cambodia, depending on rain, U.S. bombing, etc. Both bombing and trail-clearing are ΔLINKS, the one shortening, the other lengthening the range. The limiting case of a null range, incidentally, as for example with a bridge not yet built, or totally washed away, could be indicated by having the two endpoints of the range be identical.
   In effect, then, ΔLINK is the same as ΔQUAL, except that the object of the deltact is specifically a medium for a carrier rather than an object, and the property being changed is a specialized property of a medium, namely its range. The enablement and success conditions are thus the same as those for ΔQUAL.

H. Power

   8a. POWER: Several of the states, as we have seen,

---

[3] I am grateful to Fred Hornbeck for pointing out this problem to me.

require the use of various forms of "power". We group these all under the rubric POWER((A,M),(S,E)). This is the place in our system where we provide an interface between qualitative and quantitative aspects of plans. While we will not in this chapter attempt to spell out quantitative details, it is clear that they are needed for the sensible representation of mundane reality. The "harried suitor" problem (see Simon, 1972) provides a simple statement of a common type of quantitative problem in plans. The suitor is promised the favors of the starlet if he provides her with a $15,000 sports car and if he provides her with $15,000 in jewelry. The enabling condition for each gift is that the suitor have $15,000 to spend. Suppose indeed that he does have a spare $15,000. Then each act is individually enabled, but poor fellow, he cannot have the starlet. Perhaps that is just as well, she was too demanding anyway. The question is, however, would a sensible knowledge system surely recognize the type of case where resource demands should be aggregated before testing enablement conditions?

This problem pivots on whether the system can keep track of the identity or separateness of its kinds of resources. This is why we want to mark the source S of power E. We have not here taken the ultimate step of including resource gathering, or stockpiling, in the list of steps-in-plans, but it is probable the next revision of our system will include this.

8b. ΔPOWER: The fundamental idea of this deltact is that the level of power investment by the actor in a means can be increased if he has the necessary additional resources. In the notation $\Delta POWER(A,M,(S,E_1),(S,E_2),N)$, E_1 is the original power level from source S, and E_2 the final level; N is the means by which the power level is altered (e.g., putting more wood on the fire, taking more money from the bank). Means N might be a ΔHAVE, a ΔPROX, or a ΔQUAL. The first alternative covers the acquisition of an incremental power supply; the second and third, the transfer of power from another place or another form. In all three cases, the enablements and success conditions of that inner deltact become the enablements and success conditions of the ΔPOWER. Additional enablements are the definitional starting condition (a), and also (b) if N is a ΔPROX or a ΔQUAL.

(a)   POWER$((A,M),(S,E_1))$,
(b)   HAVE$((S',(E_2-E_1)),A)$--the actor must have the differential supply of power from alternative source S' (which is in a different place or from than S).

If all conditions are met, then POWER$((A,M),(S,E_2))$.

I. Units

9a. UNIT: Various pairs or sets of objects and/or actors in the world may have joint physical fates at various times for various reasons. The UNIT relation specifies what is joined to what at any given time. A unit can act as an argument in any state or deltact expression; for example, units can be objects in other units. Shirts in a suitcase form UNIT(shirts, suitcase). If a professor is carrying the suitcase, then the first unit acts as an argument[4] in the higher-order UNIT(professor, (shirts, suitcase)). When the professor boards a train carrying the suitcase, we have UNIT((professor, (shirts, suitcase)), train). It is clear that in this LISP expression mode, a unit is a list structure.

Now consider the plausible axiom that a given object cannot appear simultaneously in two disjoint units (e.g., the shirts cannot be in units with the suitcase and the drawer at the same time, unless the drawer is in the suitcase or the suitcase is in the drawer). Although this axiom is not necessary to our system, it has the interesting consequence that UNIT knowledge at any time point would consist in a partitioning of the universe of objects into disjoint list structures. The same statement could be made of PROX knowledge, incidentally (including locations as well as objects). In a sense, UNIT is a strong PROX; attachment of objects as well as proximity is required for the unit relation.

---

[4]We will follow the convention that the first argument of each unit pair is the more mobile or detachable object.

It is convenient if we denote when object Y is not in a unit with anything else by UNIT(Y)--a list consisting only of atom Y--or by UNIT(Y,NIL). Also convenient is an empty dummy statement UNIT(NIL), for reasons that will become clear later.

9b. ΔUNIT: Instances of ΔUNIT may be viewed as exchanges of elements (or units acting as elements) between two UNIT list structures. The simplest case of the formation of a unit from two disjoint objects, X and Y, can be represented by the starting conditions UNIT(X,NIL) and UNIT(Y) and the final conditions UNIT(X,Y) and UNIT(NIL). The Y and the NIL change places. This artifice using the dummy NIL is useful because we can without changing notation also express the case in which two elements originally joined become disjoined: we start with UNIT(X,Z) and UNIT(NIL), and end up with UNIT(X,NIL) and UNIT(Z). Here, Z and NIL change places. In the general case, we start with UNIT(X,Z) and UNIT(Y) and end with UNIT(X,Y) and UNIT(Z). A penny slipped from a piggy bank into a person's grasp would illustrate a change from UNIT(penny,bank) to UNIT(penny,person).

There are several different means for achieving ΔUNIT, e.g., inserting, attaching, or grasping for joining objects, and removing, detaching, or ungrasping for disjoining objects. Some of these may involve an instrument (e.g., hammer, glue, dynamite) and conceivably a power source. In effect, ΔUNIT, like ΔLINK, is a specialized form of ΔQUAL. The enablement and success conditions for ΔUNIT are therefore the same as those for ΔQUAL, except that there is a double starting condition--UNIT(X,Z) and UNIT(Y)--and there are two additional OKFORs-- OKFOR(Y,M) and OKFOR(Z,M)--which are required because Y and Z are objects rather than abstract qualities.

As noted above, the success of ΔUNIT yields UNIT(X,Y) and UNIT(Z).

IV. PLANS

A PLAN is a set of deltacts directed toward a desired

goal, these acts unfolding in series (and possibly also in parallel) under the aegis of the planner (by himself or using agents), such that earlier acts produce all the necessary but unfulfilled conditions for later acts. In Table II, the conditions are stated more formally. The notation ENABLE* includes both *enable* and *success* conditions in a single predicate. The formal statement is perhaps difficult to follow because it is recursive. To infer that an overall PLAN exists, a number of embedded sub-PLANS may be necessary. The main actor A must either himself execute an eventual D_O to cause goal S_O, or PLAN via ΔAGENCY to enlist the AGENCY of someone else (A') who will execute D_O. Further, each of the several S_i which ENABLE* D_O must either: (1) already be true; (2) be brought about by a separate PLAN of A; or (3) be brought about by a PLAN of actor A_i, whom actor A PLANs to put in AGENCY for some D_i which CAUSES S_i. It is possible for the backward ramifications of necessary PLANS to grow enormously (indeed, without limit), though often in practice most or all of the S_i in a given PLAN will be trivially satisfied in the data world without calling on further PLANS.

It was not our purpose in the specification of PLANS to give a machinery by which a clever program could construct a successful PLAN given the constraints of its particular data world--although to be sure our inferential specifications could be treated as "consequent theorems" (see Sussman, Winograd, & Charniak, 1970) and a problem-solving program turned loose on the problem of finding a route to the desired goal. The major difference in emphasis which distances us from a problem-solver orientation is that the data base can be subjective and the use of rules incomplete or flawed--we are interested, after all, not in how the planner plans, but in how an *observer* interprets the behavior of others as possibly consistent with the implementation of plans. This is what social psychologists refer to as "attribution processes" (see Jones, Kanouse, Kelley, Nisbett, Valims, & Weiner, 1974): how the causes of behavior are perceived, given access only to fragmentary data about the actor and his actions.

Such processes overlap to a great extent with artificial intelligence approaches to the drawing of inferences from limited utterances. These approaches tend to be heuristic,

TABLE II.
Specifications for a PLAN

Definitions: A=main actor
A',A_i=alternative actors as agents
S_o=goal state(X,Y)--object X
with argument Y
S_i=a subgoal state
D_i=a deltact abbreviatedly specified by *actor*,
object, and *terminus*, which CAUSES the
corresponding S_i relating *object* and
terminus.

Conventions: A line specifying PLAN, S, or D denotes that
the respective plan, state or deltact exists (is true).
PLAN has three arguments: the *actor*, the *last
deltact*, and the *goal state*. Of course PLANS include
many deltacts besides the last one; other deltacts are
represented implicitly by virtue of the following rule.

$$PLAN(A,D_o,S_o) =$$
AND(OR(D_o(A,X,Y)
AND(D_o(A',X,Y)
PLAN(A,ΔAGENCY,AGENCY(A',(D_o,A))))
)
FORALL S_i SUCH THAT ENABLE*(S_i,D_o):
OR(S_i
PLAN(A,D_i,S_i)
AND(PLAN(A_i,D_i,S_i)
PLAN(A,ΔAGENCY,AGENCY(A_i,(D_i,A))))
))

---

local, and risky. An observer or program hypothesizing the
existence of a piece of a PLAN conforming to the
specifications of Table II would not generally seek to fill in
every last detail of the complete PLAN, but would use
general plausibility arguments to sketch the main outline.
We are not in this chapter proposing how such a program

would operate. Rieger (1974) and Bruce & Schmidt (1974), from related points of view, have already developed a number of promising procedures.

V. ISSUES AND PROBLEMS

A. Omitted Concepts

Nothing has been said in this presentation about time relations. Implicitly, we have assumed time to be merely ordinal, with deltacts arranged in a specified sequence. Many deltacts by particular means, however, take substantial amounts of time, and other events can intervene during a deltact. This is more of a factor in stories of actual situations facing several interacting actors than it is in the naive plans of a single actor who does not expect complications, but it is still something of a problem. Worse from the standpoint of the planner is the coordination of two or more operations undertaken concurrently to enable later operations. Here timing may be crucial, as anyone who has ever cooked four things at once will immediately appreciate. In terms of our deltacts, some ΔQUALS have particular time courses which may carry the result beyond what is intended (e.g., OVERCOOKED rather than just COOKED). Correspondingly, certain OKFORs may be specific to a particular time window or moment.

We have also not dealt with conditionals, i.e., plans with back-up contingencies in case of failures or insufficient enablements. Neither conditionals nor timing problems seem impossible to implement within the present system, but something should clearly be done about them.

Another omission is that we have not included every major type of deltact we will eventually need in a complete system. In large measure, present omissions were motivated by the desire to postpone consideration of the complexities arising from dynamic interaction of two or more actors. We thus have not included anything in the category of "impression management"--that is, acts of social strategy aimed at the instrumental production or avoidance of certain beliefs or emotional responses in other people. Acts in this general category are of frequent occurrence in plans. Neither have we included acts of emotional response to others, although that category is not intrinsically planful, and therefore would fall under a separate type of action scheme. Even though still other deltacts may need to be added, our expectation is that the total number of deltacts need not exceed a dozen.

B. Thwarting, Negative Deltacts, and Piacts

The range of application of the deltact system can be considerably extended by inquiring into "thwarting behavior". Consider the general situation wherein actor A pursues a plan, and actor B wishes to thwart that plan. Suppose that B is aware of a particular crucial deltact D in the plan; now, what general means are available to B to intervene to his advantage? It seems that two cases can be distinguished: *prevention* and *undoing*. Actor B can inhibit deltact D from taking place by preventing or undoing one (or more) of its enabling states; he can also subvert D by preventing or undoing one (or more) of its success conditions, so that D will fail after initiation.

In Table III are informally listed representative means of preventing and undoing each of the nine states in our system. Looking at the last column first, we see that the undoing operations are (with one exception) all familiar deltacts, which simply reverse the directionality of the deltact which brought about the particular state. Stealing is a special case of ΔHAVE, spoiling is a special case of ΔOKFOR; inducing to reneg is a special case of ΔAGENCY, and so on. These we have called "negative deltacts". The one state, KNOW, which lacks a plausible negative deltact does not have the property of reversibility. Once something is known, it can never truly be not-known later (though it may be inaccessible to memory).

The other eight states, being reversible, lend themselves to the possibility of double reversal, wherein A restores the original enablement by undoing B's undoing operation. (He recovers what B took, restores the link which B severed, etc.) There is no limit in principle to the potential number of such reversals; in discourse, only the tedium of low comedy would provide a limit.

TABLE III.
Means of Thwarting the Enabling Conditions

Condition	Means of Prevention	Means of Undoing
PROX(X,Y)	Restraining X from moving	Move X from Y
HAVE(X,Y)	Keep X from Y	Take X from Y
KNOW(X,Y)	Keep X secret from Y	Brainwashing?
QUAL(X,Y)	Insulate X against Yness	Convert X from Yness
OKFOR(X,M)	Make permanent a bad quality of X	Spoil X for M
AGENCY(X,(D,Y))	Keep X from dealing with Y	Induce X to reneg on Y
LINK((U,C),(Z,Y))	Insulate terminal Y against any linking operation	Interrupt the link from X to Y
POWER((A,M),(S,E))	Corner the supply from source S	Take away or spoil a critical amount of the power supply
UNIT(X,Y)	Restrain X, lock Y, etc.	Break the connection of X and Y

When we look at the middle column of Table III, we find a rather new class of operations. Restraining an actor from moving, or keeping something a secret, or locking a door to bar entrance, etc., are all operations which forcibly preserve the status quo. Preservation is a kind of passive change, as opposed to an active change, and might be denoted by π rather than Δ. This class of projected actions we therefore call *piacts*.

It is reasonable for the planner as well as his adversary to employ piacts. If A anticipates that B intends a negative deltact (e.g., steal something A needs for the plan), he can prevent this undoing operation with a piact (e.g., keeping the critical object in a secure place). At the risk of sounding like a physicist berserk with elementary particles, it should also be noted that there must be such a thing as negative piacts, whereby the security precautions are penetrated so as to permit deltacts (either negative or positive). Whether the expanded conceptual system with two actors feverishly counteracting each other will prove to be conceptually closed (as we want it to be) is a matter requiring investigation.

C. What Kinds of Goals Are There?

One factor apparently working against the desired closedness of our PLAN structure is the seeming openness of possible goals. Human actors are capable of a bewildering variety of aspirations, from breaking the world's record for flagpole sitting to achieving spiritual salvation, or experiencing new sensual satisfactions, or making a fool out of a personal enemy, or finding a decent job, or collecting old bus transfers, and on and on and on. Our list of states includes only a few of the many possibilities (e.g., HAVE bus transfers). Conversely, many of our states do not seem like goals at all (e.g., LINK((U,C),(Z,Y)), but merely subgoals instrumental to some distant goal.

It should be recognized, therefore, that our purview in this chapter does not extend beyond the rules of chaining of mundane actions for the achievement of component goals, without specifying how the components fit together in larger thematic goals. Our PLANS, as it were, are really SUBPLANS. This is not the place to broach the categorization of thematic goals, other than to say that the major categories would very possibly resemble a list of the most important of the psychological needs distinguished by Murray (1938): Achievement, Affiliation, Dominance, etc. In this view, the goal states might be things like ACHIEVE(ACTOR X, SKILL Y) or DOMINATE (ACTOR X, UNIT Y). It would remain to articulate these goals in terms of SUBPLANS. Schank and I are presently working on this matter.

D. Episodic versus Conceptual Knowledge

Our system of states and deltacts consists of abstract enablements of abstract actions. The implicit assumption in such a system is that knowledge is applied by concretizing abstractions. For example, consider the abstract specification of OKFOR(X,M) in terms of OKFOR(Q(X),M) for all Q such that QUAL(X,Q), along with the concept that ΔOKFOR requires ΔQUAL on those qualities not suitable for M. Now let us concretize. A pipe smoker is observed to compress with his thumb the new tobacco he has just loaded into the bowl of his pipe, before attempting to light the pipe. How are we to understand this action?

The action fits the abstract OKFOR paradigm in that the tobacco is NOT-OKFOR smoking when it has the quality LOOSE, but would be OKFOR smoking with the quality PACKED. The smoker's compressing action can thus be viewed as a ΔQUAL(SMOKER, TOBACCO, LOOSE, PACKED, THUMB-PUSH) which enables OKFOR(TOBACCO, SMOKING). But need one go through this computation at all? Suppose one simply recognizes the thumb-push as a funny little thing that pipe smokers do before lighting a match, without ever thinking about why they do it? A knowledge system, informed of this action, could conceivably respond in effect, "Oh, yes. I have that sequence stored here in the pipe-smoker script." Or even, "Oh, I remember a pipe-smoker of my acquaintance who used to do that." Such a memory is *episodic* rather than propositional. Employment of the memory, in such a case, is a problem in recognition pattern-matching rather than in conceptual computation.

It is my impression that artificial intelligence workers are averse to the episodic view. It somehow seems less interesting (though it is not necessarily easier to work with, since the pattern-matching and memory size problems can be severe.) An informal measure at this point of the reader's own aversion to episodic memory theory is the extent to which he has thought up counterarguments to the pipe-smoker example (e.g., "But of course we really know *why* the pipe smoker tamps his tobacco, or at least can infer that a reason exists"). I, too, am somewhat averse to the episodic view, because the conceptual view allowing richer and more powerful system processing is more fun to work on; however, I am convinced that episodic memories are important in human information processing. To counter any counterarguments in the pipe-smoker example, therefore, I offer the following example: A slugging batsman is observed to tap his bat on home plate prior to taking his stance in the batter's box. Why is this? (Answer: Nobody knows. It's just something that many batsmen do habitually.)

It is possible that artificial intelligence workers and psychologists will part company over this issue, the former going the propositional route and the latter the episodic. Among psychologists themselves, there is much debate on this topic, with the episodic view gradually gaining ground (see Tulving, 1972). It certainly seems possible to build propositional world knowledge systems, so that even if they are not psychologically veridical (because they use abstract rules where people would use concrete past episodes or episode clusters), they are nevertheless smart.

I personally hope that such a division of orientation does not come about. As I remarked at the outset of the chapter, I find the present convergence of disciplines encouraging. I hope that ways can be found to mesh episodic and conceptual knowledge in a single system. Possibly the current interest in "frames" (Minsky, 1974; Winograd, Chapter 7) offers such a possibility.

In terms of my own theoretical system, I see two points at which episodic information may conjoin with conceptual structures, one low-level, the other high-level. At the low level, the kind of information cluster we have been suggesting in our little examples might be designated a *vignette*. Details of a prototypical scene are recruited by one or more criterial features ("Mabel saw the ants starting to crawl onto the picnic table."). If vignettes invoke scripts, they are of another type than the planning scripts in the body of this chapter. At a high level, the kind of simplified ideological superstructures which originally engaged our interest in this area have a character very similar to vignettes. The vision of Communists so imminently subverting all free nations as to require immediate counteraction is (on a broader time and space scale) metaphorically not unlike such visions as ants attacking picnics. Sitting below and above such vignette-like structures in human belief systems, however, are many conceptual structures which do not lend themselves to "scenic" representation. This chapter has been addressed to one such type of structure, the PLAN.

ACKNOWLEDGMENT

Work on this paper was facilitated by National Science Foundation Grant GS-35786.

REFERENCES

Abelson, R. P. Psychological implication. In R. P. Abelson, E. Aronson, W. J. McGuire, T. M. Newcomb, P. H. Tannenbaum & M. J. Rosenberg (Eds.), *Theories of cognitive consistency: A sourcebook.* Chicago, Ill.: Rand-McNally, 1968.

Abelson, R. P. The structure of belief systems. In R. C. Schank & K. M. Colby (Eds.), *Computer models of thought and language.* San Francisco, Ca.: Freeman, 1973.

Abelson, R. P., & Reich, C. M. Implicational molecules: A method for extracting meaning from input sentences. D. E. Walker & L. M. Norton (Eds.), *Proceedings of the First International Joint Conference on Artificial Intelligence,* 1969.

Boden, M. A. The structure of intentions. *Journal for the Theory of Social Behaviour,* 1973, *3,* 23-45.

Bruce, B., & Schmidt, C. F. Episode understanding and belief guided parsing. Presented at the meeting of the Association for Computational Linguistics, Amherst, Massachusetts, July 1974.

Charniak, E. Towards a model of children's story comprehension (AI TR-266). Cambridge, Mass.: Massachusetts Institute of Technology, 1972.

Collins, A. M., & Quillian, M. R. Retrieval time from semantic memory. *Journal of Verbal Learning and Verbal Behavior,* 1969, *8.*

Danto, A. Basic actions. *American Philosophical Quarterly,* 1965, *2,* 141-148.

Goldman, N. M. Computer generation of natural language from a deep conceptual base (Working paper 1974-2). Castagnola, Switzerland: Instituto per gli Studi Semantici e Cognitivi, 1974.

Heider, F. Attitudes and cognitive organization. *Journal of Psychology,* 1946, *21,* 107-112.

Heider, F. The role of units in social perception. In F. Heider & G. Heider (Eds.), *Social cognition.* In preparation, 1975.

Jones, E. E., Kanouse, D. E., Kelley, H. H., Nisbett, R. E., Valins, S., & Weiner, B. *Attribution: perceiving the causes of behavior.* General Learning Press, 1971.

Minsky, M. A framework for representing knowledge. In Winston, P. (Ed.), *The psychology of computer vision.* New York: McGraw-Hill, 1975.

Murray, H. A. *Explorations in personality.* London and New York: Oxford University Press, 1938.

Rieger, C. Conceptual memory. In R. Schank (Ed.), *Conceptual information processing.* Amsterdam: North-Holland, 1975.

Riesbeck, C. K. Computational understanding: Analysis of sentences and context, unpublished doctoral dissertation, Stanford University, 1974. Reprinted in R. Schank (Ed.), *Conceptual Information Processing*, 1974.

Rumelhart, D., Lindsay, P., & Norman, D. A process model for long-term memory. In E. Tulving & W. Donaldson (Eds.), *Organization of memory.* New York: Academic Press, 1972.

Schank, R. C. Conceptual dependency: A theory of natural language understanding. *Cognitive Psychology,* 1972, *3*, 552-631.

Schank, R. C. The fourteen primitive actions and their inferences (AIM-183). Stanford, California: Stanford University, Computer Science Department, 1973.

Schank, R.C. *Conceptual information processing.* Amsterdam: North-Holland, 1975.

Schank, R. C. & Abelson, R. P. Scripts, Plans and Knowledge. *Proceedings of the Fourth International Conference on Artificial Intelligence,* 1975

Schank, R. C., Goldman, N. M., Rieger, C., & Riesbeck, C. MARGIE: Memory, analysis, response generation, and inference in English. *Proceedings of the Third International Joint Conference on Artificial Intelligence,* 1973.

Simon, H. A. On reasoning about actions. In H. A. Simon & L. Siklossy (Eds.), *Representation and meaning: experiments with information processing systems.* Englewood Cliffs, N. J.: Prentice-Hall, 1972.

Sussman, G., Winograd, T., & Charniak, E. Micro-planner reference manual (AIM-203). Cambridge, Mass.: Massacusetts Institute of Technology, 1970.

Tulving, E. Episodic and semantic memory. In E. Tulving & W. Donaldson (Eds.), *Organization of memory.* New York: Academic Press, 1972.

Winograd, T. *Understanding natural language.* New York: Academic Press, 1972.

MULTIPLE REPRESENTATIONS OF KNOWLEDGE FOR TUTORIAL REASONING

John Seely Brown and **Richard R. Burton**
Bolt Beranek and Newman
Cambridge, Massachusetts

I. INTRODUCTION

This chapter provides an overview of SOPHIE, an "intelligent" instructional system which reflects a major attempt to extend Carbonell's notion of mixed-initiative Computer Aided Instruction [introduced in SCHOLAR (Carbonell, 1970)] for the purpose of encouraging a wider range of student initiatives. Unlike previous AI-CAI systems which attempt to mimic the roles of a human teacher, SOPHIE tries to create a "reactive" environment in which the student learns by trying out his ideas rather

311

than by instruction. To this end, SOPHIE incorporates a "strong" model of its knowledge domain along with numerous heuristic strategies for answering a student's questions, providing him with critiques of his current solution paths and generating alternative theories to his current hypotheses. In essence, SOPHIE enables a student to have a one-to-one relationship with an "expert" who helps the student create, experiment with, and debug his own ideas.

SOPHIEs expertise is derived from an efficient and powerful inferencing scheme that uses multiple representations of knowledge including (i) simulation models of its microcosm, (ii) procedural specialists which contain logical skills and heuristic strategies for using these models, and (iii) semantic nets for encoding time-invariant factual knowledge. The power and generality of SOPHIE stems, in part, from the synergism obtained by focusing the diverse capabilities of the procedural specialists on the "intelligent" manipulation, execution, and interpretation of its simulation models. In this respect SOPHIE represents a departure from current inferencing paradigms (of either a procedural or declarative nature) which use a uniform representation of information.

Before delving into any details about SOPHIE, we first present the basic scenario which shaped SOPHIEs outward appearance and which defined the kinds of logical and linguistic tasks it had to be able to perform. We then provide an annotated example of a student using SOPHIE followed by a discussion of its natural language processor. SOPHIEs language processor is still in its infancy; its primary interest lies in the use of a semantic "grammar" to successfully cope with the nasty problems of anaphoric references, deletions and complex ellipses inherent in any realistic man-machine dialog. We then describe the specialized inferencing techniques and the multiple representations of knowledge embodied in SOPHIE.

A. Basic Scenario

In the basic scenario, SOPHIE acts as an electronics lab instructor who helps the student transform his classroom knowledge of electronics into an experiential, intuitive knowledge of its meaning and application. It does this by interacting with the student while he is debugging a malfunctioning piece of equipment.[1] The student can perform any sequence of measurements, ask either specific questions about the implications of these measurements or more general hypothetical questions, and even ask for advice about what to consider next, given what he has discovered thus far. At any time SOPHIE may encourage the student to make a guess as to what he thinks might be wrong given the measurements he has made thus far. If he does, SOPHIE will evaluate his hypothesis by taking into consideration all the information he should have been able to derive from his current set of measurements. If any of this information is logically contradicted by the hypothesis, SOPHIE identifies and explains these contradictions. Likewise SOPHIE can judge the merits of any particular measurement with respect to the prior sequence of measurements he has made. For example, his new measurement may be logically redundant in the sense that no new information can possibly be derived from it (an extremely complex task to determine). SOPHIE can also decide if this measurement performs a reasonable split of the hypotheses space of possible faults which have not yet been ruled out by prior measurements.

It should be noted that the preceeding scenario contains quite a variety of logical tasks (i.e., hypothesis evaluation, hypothesis formation, redundancy checking, hypothetical question answering) each one of which requires a substantial amount of deep logical inferencing. One of the basic challenges in constructing SOPHIE was creating an inference system which could perform this wide range of tasks efficiently (so that it could be used in real time) and at the same time have it be robust in the sense of handling all realistic queries.

---

[1]Although throughout this chapter the domain of knowledge under consideration is electronics, the reasoning and linguistic paradigms underlying SOPHIE are applicable to many domains outside of electronics.

B. Protocol

The following protocol reveals some of the linguistic and logical capabilities of our current version of SOPHIE. We have included numerous annotations in the protocol thereby (hopefully) making it understandable to those readers not grounded in electronic jargon. In fact, simply from a linguistic point of view it is important to point out where many subtle problems arise due to anaphora and ellipsis, where domain specific "default" assignments must be made and yet might go unnoticed by a reader unfamiliar with electronics.

WELCOME TO SOPHIE, A REACTIVE LEARNING ENVIRONMENT. THE PIECE OF EQUIPMENT YOU ARE STUDYING IS A REGULATED POWER SUPPLY.[2] WOULD YOU LIKE TO START WITH A FAULTED INSTRUMENT WHICH NEEDS FIXING?
>>*YES*[3]

SOPHIE randomly selects a fault, inserts it into a model of the instrument and tells the student how the controls are set.

THE INSTRUMENT HAS A FAULT AND ITS PRESENT CONTROL SETTINGS ARE:

CC	1.0	CURRENT CONTROL
CR	HIGH	CURRENT RANGE SWITCH
LOAD	20	LOAD RESISTANCE IN OHMS
VC	1.0	VOLTAGE CONTROL
VR	HIGH	VOLTAGE RANGE SWITCH

It is assumed that the student has already had an introduction to the basic principles of the power supply so

---

[2]The schematic is shown in Fig. 1.
[3]The student's typing is shown in bold italics.

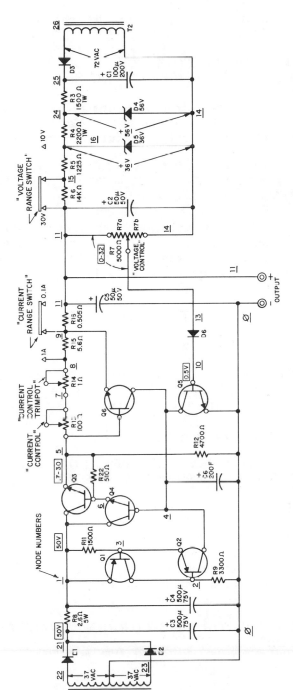

Fig. 1. Schematic diagram of circuit. The circuit is based the Heathkit IP-28 power supply which is a reasonably sophisticated power supply with both current limiting and voltage limiting behavior. These two interrelated feedback loops greatly complicate the troubleshooting process.

the above comments are sufficient to characterize the controls.

>>*WHAT IS THE OUTPUT?*
THE OUTPUT VOLTAGE IS 11.7 VOLTS.

The student begins his troubleshooting by performing measurements in the faulted circuit. Notice that "the output" was disambiguated as the output voltage.

>>*WHAT IS IT IN A WORKING INSTRUMENT?*
IN A WORKING CIRCUIT THE OUTPUT VOLTAGE IS 19.9 VOLTS.

In case the student does not know what the value of a measurement should have been, he can have any measurement repeated in a working instrument. Note the student's use of the pronoun "it" to refer to the previous measurement.

>>*WHAT IS THE VOLTAGE BETWEEN R8 AND Q1?*
THE VOLTAGE AT N1 IS 47.6 VOLTS.

The student requests an internal measurement. Notice that even though the sentence form is "voltage between x and y", it really means "voltage at z with respect to ground." The output routines reduce the input to its simpler form (i.e., between x and y reduces to Node 1).

>>*WHAT IS THE BSE EMITTER VOLTAGE OF THE VOLTAGE LIMITING TRANSITOR?*
=BASE
=TRANSISTOR
THE BASE EMITTER VOLTAGE OF Q5 IS -.037 MILLIVOLTS.

The spelling corrector has detected and corrected two misspellings. The system must also know that Q5 is the voltage limiting transistor.

>>*WHAT ABOUT FOR THE CURRENT LIMITING TRANSISTOR?*

THE BASE EMITTER VOLTAGE OF Q6 IS .349 VOLTS.

The meaning of this question is completely dependent on the context established by the prior sentence. The system uses this context to determine that he is interested in not only voltage measurements but the voltage across a particular junction.

>>*WHAT SHOULD IT BE?*
IN A WORKING CIRCUIT THE BASE EMITTER VOLTAGE OF Q6 IS .593 VOLTS.

Another way of requesting that the prior measurement be repeated in a working instrument, i.e., the meaning of "should" implies that a working instrument be used.

>>*REPLACE Q6*
I AM GOING TO ASK YOU SOME QUESTIONS ABOUT HOW Q6 IS FAULTED.

The student believes he has discovered the problem and tries to fix it. Before a component is replaced, however, the system queries him as to how he thinks it is faulted. In the case of transistors (which have complex fault modes), if he gets at least something right, the system tells him what was actually wrong and replaces the broken transistor.

ARE ANY JUNCTIONS SHORTED?
>>*I DON'T KNOW*
ARE ANY TERMINALS OPEN?
>>*YES*
WHICH TERMINAL(S) ARE OPEN?
>>*THE BASE*
Q6 IS NOT FAULTED IN THAT MANNER. Q6 NOT REPLACED.

After learning that Q6 does not have an open base, the student asks if this was at least reasonable hypothesis.

>>*IS IT POSSIBLE THAT Q6 HAS AN OPEN BASE?*

Before the system's evaluation procedures can be applied to his idea or hypothesis, the hypothesis must be completely specified. To help the student out, the system suggests a plausible complete specification of it.

>   IS THE COMPLETE FAULT MODE EB/EC/BC =
>   OP/OK/OP WHAT YOU INTENDED?
>   >>*SURE*

The fault mode should be read as: the emitter base junction (EB) is open (OP), the emitter collector junction (EC) is okay (OK) and the base collector junction (BC) is open (OP).

>   WHEN THE SETTINGS WERE:
>   CC=1.0, CR=HIGH, LOAD=20, VC=1.0, VR=HIGH
>   THAT FAULT CONFLICTS WITH THE FOLLOWING
>   MEASUREMENT.
>   YOU OBSERVED THAT THE OUTPUT VOLTAGE
>   WAS 11.7 VOLTS.
>   IF Q6 HAD FAULT MODE EB/EC/BC = OP/OK/OP
>   IT WOULD BE 23.9 VOLTS.
>   IN A WORKING CIRCUIT IT IS 19.9 VOLTS.

In determining the logical consistency of his hypothesis with respect to the measurements he has already made, a glaring counterexample is discovered.

>   ALSO:
>   THAT FAULT CONFLICTS WITH THE FOLLOWING
>   MEASUREMENT.
>   YOU OBSERVED THAT THE BASE EMITTER
>   VOLTAGE OF Q6 WAS .349 VOLTS.
>   IF Q6 HAD FAULT MODE EB/EC/BC = OP/OK/OP
>   IT WOULD BE .712 VOLTS.
>   IN A WORKING CIRCUIT IT IS .593 VOLTS.

The system then points out a more subtle counterexample.

>   THAT FAULT DOES NOT EXPLAIN ANY OF THE
>   SYMPTOMS YOU OBSERVED BUT IT IS
>   CONSISTENT WITH THE VALUES FOR THE BASE

EMITTER VOLTAGE OF Q5 AND THE VOLTAGE AT N1.

Finally the system tells him what is right or at least consistent with his hypothesis.

>>*WHAT IS THE OUTPUT OF THE CONSTANT CURRENT SOURCE?*
THE OUTPUT OF THE CONSTANT CURRENT SOURCE IS -.59 MILLIAMPS.

To understand this request the system must also know about the functional blocks of the instrument. Also note that the system had to determine that "output" in this context meant "output current" as opposed to "output voltage" as was meant in the first question of the protocol!

>>*V BETWEEN N4 AND N5*
THE VOLTAGE BETWEEN N4 AND N5 IS 1.64 VOLTS.

Noun phrase utterances get interpreted as questions. "V" is an accepted abbreviation for voltage.

>>*COULD THE PROBLEM BE THAT THE BASE OF Q3 IS OPEN?*

The student constructs another hypothesis

IS THE COMPLETE FAULT MODE EB/EC/BC = OP/OK/OP WHAT YOU INTENDED?
>>*YES*
UNDER THE SETTINGS:
CC=1.0, CR=HIGH, LOAD=20, VC=1.0, VR=HIGH
THE FAULT WHICH YOU PROPOSED WILL PROPAGATE.
LIKELY FAULTS WHICH MIGHT OCCUR ARE:
R22 OPEN
WOULD YOU LIKE ME TO CONTINUE TESTING YOUR HYPOTHESIS IN A CIRCUIT IN WHICH R22 IS OPEN ALSO?
>>*NO*

The hypothesis proposed by the student logically entails another component being blown. Before further evaluation of his hypothesis, the user is queried as to whether he realizes this entailment. If he does not, he is given the chance to have his hypothesis evaluated in a "ghost" world in which that component would not have blown.

WOULD YOU LIKE ME TO COMMENT ON YOUR HYPOTHESIS AS IF NO OTHER PARTS HAD BLOWN?
>>YES
WHEN THE SETTINGS WERE:
CC=1.0, CR=HIGH, LOAD=20, VC=1.0, VR=HIGH,
THAT FAULT WOULD HAVE CAUSED THE FOLLOWING SYMPTOM.
YOU OBSERVED THAT THE VOLTAGE BETWEEN N4 AND N5 WAS 1.64 VOLTS.
IF Q3 HAD FAULT MODE EB/EC/BC = OP/OK/OP IT WOULD BE 31.3 VOLTS.
IN A WORKING CIRCUIT IT IS 1.7 VOLTS.

ALSO:
THAT FAULT CONFLICTS WITH THE FOLLOWING MEASUREMENT.
YOU OBSERVED THAT THE OUTPUT VOLTAGE WAS 11.7 VOLTS.
IF Q3 HAD FAULT MODE EB/EC/BC = OP/OK/OP IT WOULD BE 1.19 VOLTS.
IN A WORKING CIRCUIT IT IS 19.9 VOLTS.

THAT FAULT DOES NOT EXPLAIN ANY OF THE SYMPTOMS YOU OBSERVED BUT IT IS CONSISTENT WITH THE VALUES FOR THE OUTPUT CURRENT OF THE CONSTANT CURRENT SOURCE, THE BASE EMITTER VOLTAGE OF Q5 AND THE VOLTAGE AT N1.

Again the student's hypothesis fails.

>>SAVE
CONTEXT SAVED, TO RETURN TO IT TYPE 'RESTORE'.

YOU NOW HAVE A WORKING INSTRUMENT WITH WHICH TO EXPERIMENT.

The student has discovered more flaws in his understanding and is probably surprised by the high voltage between N4 and N5 that follows from his hypothesis (i.e., 31.3 volts). Wishing perhaps to explore what caused R22 to blow and what led to this unexpectedly high voltage he requests that the present troubleshooting context be saved (i.e., his fault and all his measurements which are the gist for the evaluation specialist). He can now perform any experiments he wishes.

>>*WHAT IS THE CURRENT THRU R22 WHEN IT IS SHORTED?*
IF R22 WERE SHORTED UNDER THE PRESENT SETTINGS, THE CURRENT THRU R22 WOULD BE 61 MILLIAMPS.
IN A WORKING CIRCUIT IT IS 1.8 MILLIAMPS.

This is an example of a hypothetical or conditional question. Note that "it" refers to R22 which is embedded in a noun phrase earlier in the same statement.

>>*WHAT ARE THE SPECS OF R22?*
THE SPECIFICATIONS OF R22 ARE:
VALUE -- 510 OHMS
POWER/RATING -- .5 WATTS

He now asks a factual question which simply gets looked up in the semantic net.

>>*RESTORE*
PREVIOUS SITUATION RESTORED.

He may have figured out why R22 blew when Q3 opened and now wishes to get back to the task of troubleshooting.

>>*WHAT COULD BE WRONG?*
HERE ARE SOME FAULTS WHICH EXPLAIN THE MEASUREMENTS YOU OBSERVED:
THE BETA OF Q3 IS LOW
THE BETA OF Q4 IS LOW

OR R22 HAS THE WRONG RESISTANCE

He seems to have run out of ideas and requests some help. At this point the hypothesis generation "specialists" are invoked. These procedures examine all the measurements he has thus far made and then construct a collection of hypotheses which are logically consistent with the values of these measurements.

>>*IS THE EMITTER CURRENT OF Q4 CORRECT?*
THE EMITTER CURRENT OF Q4 IS -61 MILLIAMPS.
IN A WORKING CIRCUIT THE EMITTER CURRENT OF Q4 IS -49 MILLIAMPS.
THAT VALUE SEEMS TO BE INCORRECT.

This question is in some sense a cheat. It really consists of two measurements. If SOPHIE were assigning costs to his measurements or questions he would be charged accordingly.

>>*COULD R22 BE LOW?*
IS IT OKAY IF I USE 25 OHMS FOR THE RESISTANCE OF R22?
>>*YES*
THAT FAULT AGREES WITH ALL OF THE MEASUREMENTS YOU HAVE MADE.

Finally he generates a reasonable hypothesis, but this does not mean that R22 is necessarily low! Just that R22 being low is consistent with what he should be able to determine from his measurements thus far.

>>*WHAT IS THE RESISTANCE OF R22?*
THE MEASURED RESISTANCE OF R22 IS 25 OHMS.
>>*REPLACE R22*
R22 REPLACED
NOW YOU SHOULD TAKE MEASUREMENTS TO SEE IF THERE IS ANYTHING ELSE WRONG.

The system decides he must know what is wrong with R22 so it does not hassle him like it did when he asked for Q6 to be replaced.

>>*NOW WHAT IS THE OUTPUT?*
THE OUTPUT VOLTAGE IS 19.9 VOLTS.
>>*WITH A 1000 OHM LOAD?*
WITH THE LOAD NOW SET TO 1000
THE OUTPUT VOLTAGE IS 30 VOLTS.

He discovers that the unit is now functioning correctly. Note the ability of the system to handle rather extensive ellipses such as "with a 1000 ohm load".

>>*WHAT   WOULD   HAPPEN   IF   R22   WERE SHORTED?*
IF R22 WERE SHORTED UNDER THE PRESENT SETTINGS:
CC=1.0, CR=HIGH, LOAD=1000, VC=1.0, VR=HIGH
THERE WOULD BE NO CHANGE IN THE OUTPUT VOLTAGE (I.E. 30 VOLTS).
HOWEVER UNDER THE SETTINGS:
CC=1.0, CR=HIGH, LOAD-10, VC=1.0, VR=LOW,
THE OUTPUT VOLTAGE WOULD BE .623 VOLTS.
IN A WORKING CIRCUIT IT IS 9.93 VOLTS.

A good example of what else is involved in handling hypothetical questions. Here the system first guesses that he is interested in output voltage since he didn't specify anything particular. Then it fails to find any change of the output voltage with the present load resistance. However, before it concludes that R22 has no effect it uses some other knowledge about the circuit to decide that a better test case or example would be to set the instrument up with a heavier load. Under this setting it does find a significant change.

>>*GOODBYE*

II. NATURAL LANGUAGE UNDERSTANDING

Since we are trying to build an environment in which students will create and articulate ideas, we must provide a convenient way for them to communicate these ideas to the system. A student will become very frustrated if, after forming an idea, he has to try several ways of expressing it before the system understands it. In addition he will become bored if there is a long delay (say 10 seconds) before he gets a response. To compound our problem we discovered from using early versions of SOPHIE that when a person communicates with a logically "intelligent" system he inevitably starts to assume that the system shares his "world-view" or is at least "intelligent" in the linguistic art of following a dialog. In other words, SOPHIE had to cope with problems such as anaphoric references, context-dependent deletions, and ellipses which occur naturally in dialogs. In fact handling these constructs seemed more important than building a system endowed with great syntactic paraphrase capabilities.

A. Semantic Grammar

These realizations led us to build a natural language processor based on the concept of a semantic grammar. A semantic grammar captures some of the simpler notions of conceptual dependencies, enabling one to predict on semantic grounds the referent of an anaphoric expression or the element which has been ellipsed or deleted. In a semantic grammar the usual syntactic categories such as noun, noun phrase, verb phrase, etc. are replaced by semantically meaningful categories. These semantic categories represent conceptual entities known to the system such as "measurements", "circuit elements", "transistors", "hypotheses", etc. (While such refinement can lead to a phenomenal proliferation of nonterminal categories in a grammar, the actual number is limited by the number of underlying concepts which can be discussed. For SOPHIEs present domain, there are on the order of 50 such concepts.)

The grammar which results from this refinement is a

formal specification of constraints between concepts. That is, for each concept there is a grammar rule which explicates the ways of expressing that concept in terms of its constituent concepts. Each rule also provides explicit information concerning which of its constituent concepts can be deleted or pronominalized. Once the dependencies have formalized into the semantic grammar, each rule in the grammar is encoded (by hand) as a LISP procedure. This encoding process imparts to the grammar a top-down control structure and specifies the order of application of the various alternatives of each rule. The resulting collection of LISP functions constitute a goal-oriented parser in a fashion similar to SHRDLU (Winograd, 1972).

Encoding the grammar as LISP procedures shares many of the advantages which ATNs[4] (Woods, 1970) have over using traditional phrase structure grammar representations. Four of these advantages are:

(i) the ability to collapse common parts of a grammar rule while still maintaining the perspicuity of the semantic grammar,

(ii) the ability to collapse similar rules by passing arguments (as with SENDR),

(iii) the ease of interfacing other types of knowledge (in SOPHIE, primarily the semantic network) into the parsing process, and

(iv) the ability to build and save arbitrary structures during the parsing process.

In addition to the advantages it shares with ATN representation, the LISP encoding has the computational advantage of being compilable directly into efficient machine code.

Result of the Parsing: Basing the grammar on conceptual entities eliminates the need for a separate semantic interpretation phase. Since each of the nonterminal categories in the grammar is based on a semantic unit, each

---

[4]All of these advantages are, of course, also shared by a PROGRAMMAR grammar (Winograd, 1972).

rule can specify the semantic description of a phrase that it recognizes in much the same way that a syntactic grammar specifies a syntactic description. Since the rules are encoded procedurally, each rule has the freedom to decide how the semantic descriptions returned by the constituent items of that rule are to be put together to form the correct "meaning".

For example, the meaning of the phrase "Q5" is just Q5. The meaning of the phrase "the collector of Q5" is (COLLECTOR Q5) where COLLECTOR is a function encoding the meaning of "collector". "The voltage at the collector of Q5" becomes (MEASURE VOLTAGE (COLLECTOR Q5)) where MEASURE is the procedural specialist who knows about the concept of a measurement. The relationship between a phrase and its meaning can be straightforward and, if the concepts and the specialists in the query language are well matched, usually is. It can get complicated, however. Consider the phrases "the base emitter of Q5 shorted" and "the base of Q5 shorted to the emitter". The thing which is "shorted" in both of these phrases is the "base emitter junction of Q5." The rule which recognizes both of these phrases, PART/FAULT/SPEC, can handle the first phrase by invoking its constituent concepts of JUNCTION (base emitter of Q5) and FAULT/TYPE (shorted) and combining their results. In the second phrase, however, it must construct the proper junction from the separate occurrences of the two terminals involved. Notice that the parser does some paraphrasing, as the "meaning" of the two phrases is the same.

The result returned by the parser is the "meaning" of the entire statement in terms of a simple program. This program specifies which of the procedural specialists should be called (and in what order) to calculate an answer to the student's question or perform the student's command. It is also used by the output generation routines to construct an appropriate phrasing of the response.

B. Use of Semantic Information During Parsing

Prediction: Having described the notion of a semantic grammar, we now describe the ways it allows semantic information to be used in the understanding process. One use of semantic grammars is to predict the possible alternatives that must be checked at a given point. Consider for example the phrase "the voltage at xxx" (e.g., "the voltage at the junction of the current limiting section and the voltage reference source"). After the word "at" is reached in the top-down, left-right parse, the grammar rule corresponding to the concept "measurement" can predict very specifically the conceptual nature of "xxx", i.e., it must be a phrase specifying a location (node) in the circuit.

This predictive information is also used to aid in the determination of referents for pronouns. If the above phrase were "the voltage at it", the grammar would be able to restrict the class of the possible referents to locations. By taking advantage of the available sentence context to predict the semantic class of possible referents, the referent determination process is greatly simplified. For example:

    (1a) Set the voltage control to .8?
    (1b) What is the current thru R9?
    (1c) What is it with it set to .9?

In (1c), the grammar is able to recognize that the first "it" refers to a measurement (that the student would like re-taken under slightly different conditions). The grammar can also decide that the second "it" refers to either a potentiometer or to the load resistance (i.e., one of those things which can be set.). The referent for the first "it" is the measurement taken in (1b), the current thru R9. The referent for the second "it" is "the voltage control" which is an instance of a potentiometer. The context mechanism which selects the referents will be discussed later.

Simple Deletion: The semantic grammar is also used to recognize simple deletions. The grammar rule for each conceptual entity knows the nature of that entity's constituent concepts. When a rule cannot find a constituent concept, it can either

    (i) fail (if the missing concept is considered to be obligatory in the surface structure representation), or
    (ii) hypothesize that a deletion has occurred and continue.

For example, the concept of a TERMINAL has (as one of its realizations) the constituent concepts of a TERMINAL-TYPE and a PART. When its grammar rule only finds the phrase "the collector", it uses this information to posit that a part has been deleted (i.e., TERMINAL-TYPE gets instantiated to "the collector" but nothing gets instantiated to PART). SOPHIE then uses the dependencies between the constituent concepts to determine that the deleted PART must be a TRANSISTOR.

Ellipses: Another use of the semantic grammar allows the processor to accept elliptic utterances. These are utterances which do not express complete thoughts (i.e., a completely specified question or command) but only give differences between the underlying thought and an earlier one.[5] For example, (2b) and (2c) are elliptic utterances.

(2a) What is the voltage at Node 5?
(2b) At Node 1?
(2c) What about between nodes 7 and 8?

There is a grammar rule for elliptic phrases which is aware of which constructs are frequently used to contrast similar complete thoughts and recognizes occurrences of these as ellipses. This grammar rule identifies which concept or class of concepts are possible from the context available in a elliptic utterance. Later we will discuss the mechanism that determines to which complete thought an ellipsis refers.

C. Using Context to Determine Referents

Pronouns and Deletions: Once the parser has determined the existence and class (or set of classes) of a pronoun or deleted object, the context mechanism is invoked to determine the proper referent. This mechanism has a history of student interactions during the current session which contains for each interaction the parse (meaning) of the student's statement and the response calculated by the

---

[5]This is not strictly the standard use of the word "ellipsis."

system. This history list provides the range of possible referents and is searched in reverse order to find an object of the proper semantic class (or one of the proper classes). To aid in the search of the history list, the context mechanism knows how each of the procedural specialists which can appear in a parse uses its arguments. For example, the specialist MEASURE has a first argument which must be a quantity and a second argument which must be a part, a junction, a section, a terminal, or a node. Thus when the context mechanism is looking for a referent which can either be a PART or a JUNCTION, it will look at the second argument of a call to MEASURE but not the first. Using the information about the specialists, the context mechanism looks in the present parse and then in the next most recent parse, etc. until an object from one of the specified classes is found.

The significance of using the specialist to filter the search instead of just keeping a list of previously mentioned objects is that it avoids misinterpretations due to object-concept ambiguity. For example, the object Q2 is both a part and a transistor. If the context mechanism is looking for a part, Q2 will be found only in those sentences in which it is used as a part and not in those in which it is used as a transistor. In this way the context mechanism finds the most recent occurrence of an object being used as a member of one of the recognized classes.

Referents for Ellipses: If the problem of pronoun resolution is looked on as finding a previously mentioned object for a currently specified use, the problem of ellipsis can be thought of as finding a previously mentioned use for a currently specified object. For example,

(3a) What is the base current of Q4?
(3b) In Q5?

The given object is "Q5" and the earlier function is "base current". For a given elliptic phrase, the semantic grammar identifies the concept (or class of concepts) involved. In (3b), since Q5 is a transistor, this would be TRANSISTOR. The context mechanism then searches the history list for a specialist in a previous parse which accepts the given class as an argument. When one is found, the new phrase is substituted into the proper argument position and the substituted meaning is used as the meaning of the ellipsis. Currently recognition of ellipsed information proceeds in a top-down fashion. In a domain which has many possible concepts used in ellipses, the recognition of the ellipsed concept should either proceed bottom-up or be restricted to concepts recently mentioned.

D. Fuzziness

Having the grammar centered around semantic categories allows the parser to be sloppy about the actual words it finds in the statement. This sense of having a concept in mind, and being willing to ignore words to find it, is the essence of keyword parsing schemes. It is effective in those cases where the words that have been skipped are either redundant or specify gradations of an idea which are not distinguished by the system. Semantic grammars provide the ability to blend keyword parsing of those concepts which are amenable to it with the structural parsing required by more complex concepts.

The amount of sloppiness (i.e., how many (if any) words in a row can be ignored) is controlled in two ways. First, whenever a grammar rule is invoked, the calling rule has the option of limiting the number of words that can be skipped. Second, each rule can decide which of its constituent pieces or words are required and how tightly controlled the search for them should be. Taken together, these controls have the effect that the normal mode of operation of the parser is tight in the beginning of a sentence but more fuzzy after it has made sense out of something.

E. Results

Our two goals for SOPHIEs natural language processor are efficiency and friendliness. In terms of efficiency, the parser has succeeded admirably. The grammar written in INTERLISP (Teitelman, 1974) can be block compiled.

Using this technique, the complete parser takes up about 5k of storage and parses a typical student statement consisting of 8 to 12 words in around 150 milliseconds! Our goal of friendliness is much harder to measure since the only truly meaningful evaluation must be made when students begin using SOPHIE in the classroom. Our results so far, however, have been encouraging. The system has been used in hundreds of hours of tests by people involved in the SOPHIE project. In addition, several dozen different people have had realistic sessions (as opposed to demonstrations) with SOPHIE and the parser was able to handle most of the questions which were asked. Anytime a statement is not accepted by the parser, it is saved on a disk file. This information is constantly being used to update and extend the grammar.

F. Expanding the Natural Language Processor

Areas in which the natural language processor is lacking at present include relative clauses, quantifiers, and conjunctions -- the most noticeable being the lack of conjunction. While incorporating conjunction in a systematic way will almost certainly require an additional mechanism, the semantic nature of our nonterminal categories and the predictive ability of the semantic grammar should provide a good handle on the combinatorial explosion normally accompanying conjunction.

Another area in which the semantic grammar looks especially useful is in providing constructive feedback to the user when one of his statements fails to parse. Such feedback is very important for without it the user does not know whether to try rephrasing his question (if so, how) or to give up altogether on this line of questioning. In general, when a statement is not accepted by our top-down parser, little information is left around about why the sentence was not parsed. This is especially true if the unacceptable part occurs near the beginning of the sentence. (Our parser is working left to right.) A bottom-up parsing scheme has the advantage in this respect that constituents are recognized wherever they occur in the sentence. Combining a bottom-up parsing scheme with the semantic grammar provides a method for generating semantically meaningful feedback. After a sentence fails to parse, it can be passed through a bottom-up parser using a modified version of the semantic grammar.[6]   Since the grammar is semantically based, the constituents found in the bottom-up parse represent "islands" of meaningful phrases. The modified semantic grammar can then be looked at to discover possible ways of combining these islands. If a good match is found between one of the rules in the grammar and some of the islands, another specialist can use the grammar to generate a response which indicates what other semantic parts are required for that rule. Even if no good matches are found, a positive statement can often be made which explains the set of possible ways the recognized structures could be understood. We think such positive feedback can be critical to breaking the user out of a vicious cycle of attempting syntactic paraphrases of a semantically unrecognizable idea by providing him explicit clues as to the set of things that can be understood by the system in that local context. Mechanisms for handling the conjunction and feedback problems as well as other issues relating to semantic grammars will be discussed in a later paper (Burton, 1975).

III. ON INFERENCING

In order to put the remaining part of the chapter in perspective let us review the different kinds of logical tasks that SOPHIE must perform. First there is the relatively straightforward task of answering hypothetical questions of the form: "If X then Y", where "X" is an assertion about some component or setting of the given instrument and "Y" is a question about its resultant behavior. A simple example might be: If the base-emitter junction of the

---

[6]Actually, a simplified, non-procedural form of the semantic grammar would be used. Here is a good example of using multiple representations of knowledge: a procedural (non-introspectable) version of the grammar for top-down parsing and a simplified non-procedural version used for making comments.

voltage limiting transistor opens, then what happens to the output voltage?"

The second task involves hypothesis evaluation of the form: "Given the measurements I have thus far made could the problem be X", where X is an assertion of the state of a given component. For example: "Could the base of Q3 be open?" What is at stake here is not determining whether the assertion X is true (i.e., whether Q3 is open in the faulted circuit!), but rather determining if the assertion X is logically consistent with the information already collected by the student. If it is not consistent, then the system must demonstrate why it is not. Likewise if it is consistent, the system must identify the subset of the collected information that supports the assertion and the subset which is independent of it.

The third logical task is that of hypothesis generation. In its simplest form this task involves constructing all possible "hypotheses" (or possible worlds) that are logically consistent with the known information, i.e., consistent with the information derivable from the current set of measurements. This task can be solved by the classical "generate and test" paradigm where the "test" part of the paradigm is performed by the previously mentioned hypothesis evaluation system. The "generate" part therefore forms the heart of this system.

The final task involves the complex and subtle issue of deciding whether a given measurement could in principle add any new information to what is already known. That is, is the given measurement logically redundant with respect to previous measurements or, stated differently, could the result of this measurement have been predicted from the previous measurements and a complete theory of the circuit.

We have found that all these logical tasks can be conveniently achieved by our model-driven, example-based, inference mechanism. Since simulation is at the heart of this system, we begin our technical discussion by considering how simulation is used to answer hypothetical questions.

A. Hypothetical Questions

A drastically simplified but intuitive view of one way to handle hypotheticals such as "If R11 shorts, then does the output current drop?" is simply to try it out and see. That is, instead of trying to deduce the consequences of R11 shorting, why not short R11 in some virtual but executable model of the given circuit, and then "run" the model to see what the consequences are? There are numerous complications to such a scheme, but instead of discussing them in the abstract, let us examine how this basic paradigm is realized in SOPHIE.

Given the hypothetical question "If X, then Y?", a procedural specialist, well versed in the inner workings of SOPHIEs general purpose simulator (Nagel & Pederson, 1973; Brown, Burton, & Bell, 1974), is passed the assertion X. This specialist first determines if the assertion unambiguously specifies a modification to the circuit.[7] If it does, then the specialist modifies the circuit description residing in the simulator so as to make it consistent with the assertion X.[8] Following this operation, the simulation model is executed, thereby producing as output a voltage table which specifies for each node in the circuit its voltage with respect to ground. This voltage table contains, either explicitly or implicitly, *all* the logical consequences of this modification under the current context or boundary conditions (i.e., instrument settings, load resistance, etc.).

Because this table contains a great deal of implicit information, it is treated as a structured data base by a collection of question-answering specialists which know how

---

[7]Examples of ambiguous or underspecified modifications are: capacitor being leaky -- how leaky; terminal being opened on a transistor -- what about the other terminals; beta shift in a transistor -- how much did it shift. In such cases as these, the specialist either queries the user or makes a default assignment, depending on the context of the question.

[8]This specialist enables (nontopological) modifications of the circuit to be made without requiring the simulator to redetermine the circuit equations. Hence, the invariant aspects of the circuit get analyzed once and compiled into an efficient model.

to derive information contained only implicitly in the data base. For example, a CURRENT specialist can determine from the data base the current flowing through every component or junction in the circuit; a RESISTANCE specialist can determine the "active" resistance of any component by using Ohm's law and the output of the CURRENT specialist. There is a power dissipation specialist and so on. The point is that each of these specialists has the knowledge (and inference capabilities) to compute additional (implicit) information contained in the generated data base.

A hypothetical question is then answered by transforming the question "Y" into calls to the appropriate question-answering specialists which construct the answer from the data base. Note the flow of information in answering the "if X, then Y" type question. X is first interpreted and "simulated" thereby generating a data base (or hypothetical world state) which implicitly contains all the consequences of X. Then Y gets interpreted, resulting in a directed action to infer particular information from this data base. Notice that the data base is generated without regard to Y -- a policy which on the surface may seem wasteful, but which proves not to be, as is discussed in Chapter 4. The above exposition has admittedly overlooked certain complications. For example, as anyone who has tinkered with any kind of complex systems well knows, a proposed modification to a system can result in disaster. Electronics is no exception. A modification often entails certain unexpected side effects of components sizzling into a vapor state which must somehow be captured in handling hypothetical questions. Stated somewhat more precisely, the data base or voltage table generated by the simulator satisfies only some of the constraints which constitute a complete theory of the circuit, its components and the general laws of electronics. In particular, the voltage table satisfies all of Kirchhoff's laws and the laws defining the behavior of transistors, etc. but the simulator does not attempt to satisfy "meta" constraints such as the limited power dissipation of the components. Indeed, since simulators are usually used to simulate working or near-working circuits there is little point in checking for such violations; but for our use such checks are crucial!

The simple paradigm mentioned above must therefore be expanded to handle the case where a proposed modification causes certain components to blow, i.e., a meta-constraint violation. Briefly, several new specialist are required. The first specialist examines the data base generated by the simulator in order to infer whether any meta-constraints such as power dissipation, voltage breakdown factor, etc. have been exceeded. This task involves repeated calls on the question-answering specialist. After determining all such violations it passes them on to another specialist which decides, using some heuristic knowledge of electronics, which violation is most severe and therefore which component is most likely to blow. This specialist then translates the selected violation (e.g., a particular resistor being overloaded) into a call for an additional modification of the circuit (e.g., that resistor opens) and fires up the model accordingly. Only one modification at a time is made since often one component blowing will "protect" another component even though initially both were overloaded. This process is then repeated until a point is then reached in which the output of the simulator satisfies all the meta-constraints.

The above process has now generated two important structures. First it has generated a kernel data base for various question-answering specialists. Second, it has generated a tree of possible fault propagations in conjunction with a control path of successive calls to the simulator. This latter structure reflects a sense of causality[9] which can be used to ascertain a causal chain of "important" events which followed from a particular modification.

There remains one crucial technicality worth discussing before we move on to the more novel uses of this inference scheme. This technicality has to do with the implicit quantifiers that usually lurk behind the scenes in nearly

---

[9]Since the simulator basically uses relaxation techniques, no local sense of causality is forthcoming from one particular simulation run; however, by factoring the "theory" of the circuit into constraints and meta-constraints we get the efficiency of relaxation but at the same time a sense of causality.

all hypothetical questions. For example, consider the hypothetical question: "If R22 shorts then does the output voltage change?" At first glance handling this question would seem straightforward: simply modify the circuit description (in the simulator) to make R22 have zero resistance and then execute the simulation system and examine the output voltage. Note, however, that nowhere has there been any specification of the boundary conditions (i.e., switch settings, load resistance, etc.) under which to run the simulator. In principle there can be an infinite number of conditions to try, each of which would require an execution of the simulator.

Our solution to this predicament involves the notion of using a weak or incomplete theory of the circuit to suggest potentially "useful" boundary conditions to be tried in order to obtain answers to particular questions. For example, SOPHIE first uses the present instrument settings -- remember such questions always occur in a context. Next, if the answer has not been determined (in this case, if output voltage has not changed) a specialist would attempt to construct a set of "critical" boundary conditions.

The heuristics underlying this specialist rely on the observation that all complex circuits have a hierarchical functional decomposition. Each module in this decomposition has both a structural description of its components (and their interconnections) and also a teleological description of its purpose in the overall design of the circuit or at least with respect to its superordinate module. From these teleological descriptions, it is possible to determine a set of boundary conditions which will force the circuit into a set of states which invoke or test out the various purposes of each module. Given these test cases[10]

---

[10]SOPHIE currently cannot deduce such cases. Instead, associated with the description of each module in the semantic net is an extensional specification of test cases derived from such qualitative knowledge as "when testing the Darlington amplifier make sure it can deliver its maximum current". Hence, when putting in a new circuit to SOPHIE, not only must a circuit description and a functional block description be put into the semantic net, but also the test cases for each functional block.

then all this specialist needs to do is to determine the chain of modules which contain the component that is being hypothetically changed and use these test cases as boundary conditions for the simulator. For example, in the above case, the specialist would discover that R22 was part of the Darlington amplifier (from accessing information in the semantic net) and that this module could be stressed by using a heavy load and setting the voltage and current controls to maximum. This *reasoning-by-example* paradigm is especially power-ful in SOPHIE since its use of simulation models provides a particularly effective technique for constructing or filling out examples that meet "interesting" conditions or constraints.

Before describing how this basic technique can be generalized to handle the spectrum of other logical tasks performed by SOPHIE, we call attention to the explicit factorization of processes and the multiple representations of knowledge underlying this single logical task. There are four basic modules to this factorization. The first is the simulator or data base generator which embodies a set of constraints reflecting general laws of electronics (e.g., Kirchhoff's laws, Ohm's laws), accurate models of transistors, resistors, capacitors plus a set of constraints defining the given circuit. Executing the simulation produces a description of a "world state" which simultaneously satisfies all these constraints. This data base constitutes the second module of this factorization. It, however, is not an arbitrary collection of assertions which describe the "world state" but is instead a carefully designed modelling structure embodying a kernel set of information from which the answers to any questions about the state can be efficiently derived through the application of question-answering or measurement specialists. These inference specialists constitute the third module. They, likewise, embody general principles of electronics and use these principles (much as Consequent theorems are used in Planner) to determine information that is contained only implicitly in the data base. Since the generated data base is always of a specific form, the question-answering specialists can be designed to take advantage of this invariant structure by having all their "how to do it" type knowledge encoded in terms of how to operate on this fixed

set of kernel relations or predicates.   The significance of
this three-part factorization is discussed at greater length
in Chapter 4.

The fourth module contains qualitative knowledge (e.g.,
what components are most likely to blow, how power
amplifiers can be stressed) and heuristic strategies for
combining the qualitative knowledge with the other three
modules.   Speaking somewhat metaphorically, this fourth
module may be viewed as a "weak" or incomplete theory (of
electronics) which has been constructed for carrying out a
particular task.   Much of what follows concerns how
additional weak theories can be used to augment the
powerful but narrow capabilities of the data base generator
(and its corresponding interrogators) so that it can be used
to perform other kinds of logical tasks besides just
answering hypothetical questions.

B. Hypothesis Evaluation

Hypothesis evaluation is the process of determining the
logical consistency of a given hypothesis[11] with respect to
the information derivable from the current set of
measurements.   It is important to realize that a hypothesis
can be logically consistent with the known information and
still not be correct in the sense of specifying what is
actually wrong with the circuit.   For example, if no
measurements have been performed -- meaning that in
principle no information is known about the behavior of
the instrument -- then many hypotheses are acceptable (i.e.,
those which are syntactically consistent).   If, however, some
measurements have been made, then the task for the
hypothesis evaluation specialist is to partition these
measurements into three classes.   One class contains the
measurements that are contradicted by the given hypothesis,
another class contains the measurements which are logically
entailed by the hypothesis and the last class contains those

---

[11]A hypothesis concerns the state of a given component such as
a capacitor being shorted, a resistor being open, a transistor being
shorted, etc.

measurements that are independent of any of the logical consequences of the given hypothesis.

Although these partitions are only over the set of measurements the student has taken, they are determined by taking into consideration all the logical implications derivable from the given hypothesis. If, for example, a hypothesis concerns a particular component being shorted, there need be no direct or obvious measurement on that component for that hypothesis to be either supported or refuted! By taking into consideration both the local and global interactions of components in the circuit, measurements arbitrarily far away from the hypothetically faulted component may be used to support or refute the hypothesis.

By restricting the domain of acceptable hypotheses to statements specifying faulty components of the circuit, simulation can be used to determine the consequences of a hypothesis much as it was used to infer the consequences of the assertional part of a hypothetical question. Unlike the handling of hypothetical questions, there is no inherent problem with determining which boundary conditions to use. We simply use the same set of boundary conditions which the student used while performing his given set of measurements. (Each measurement has associated with it the complete specification of how the instrument was set up when that measurement was taken.)

To facilitate a concise description of how a student's hypothesis is evaluated, we introduce the notion of a context frame which consists of the set of measurements the student made under one particular setting of the instrument. In other words, a context frame is all those measurements made under the same boundary conditions. The hypotheses evaluation specialist proceeds as follows: First it selects a context frame (using various psychological considerations such as recency, number of measurements, etc). It then uses the boundary conditions of that context to set up a simulation of the hypothesis. The output of the simulation is then used by the question-answering or measurement specialists to reconstruct, in this generated "hypothetical world", all the measurements composing the selected context frame. If the values of any of these measurements are not equivalent to the ones taken by the student in the given frame, then a counterexample or inconsistency has been established. Depending on other considerations (of either a pedagogical or logical nature) another context frame is selected and the process is repeated. If none of the context frames yields a contradiction, then the hypothesis is accepted as being logically consistent with all the known information.

There remain two unresolved issues. First we have not specified how to separate those measurements supported by the hypothesis from those which are independent of it. Second, and much more important, we have relied exclusively on the quantitative replication of the values of these measurements in the hypothetical world (i.e., the world entailed by the hypothesis) with those actually obtained by the student. This is a most precarious strategy, for few people can construct hypotheses that exactly mimic the quantitative behavior they have thus far observed, and furthermore, there is no reason why they should be able to! What is reasonable to expect is a more qualitative, common sense mimicry of the results in the observed world by those in the hypothetical world. In order to determine this, a "metric" is used to decide if the two exact quantitative values of a measurement (each performed in its "world") are *qualitatively* similar. For example are .3 and .9 "equivalent" values for the voltage at some node? In principle, this metric must incorporate both a general theory of electronics (such as the expected voltage range of a forward biased base-emitter junction) plus a structural theory of the particular circuit.

Our solution to this problem employs a heuristic which circumvents much (but not all) of the need for employing such theories. It is based on the observation that the value of a given measurement in a working circuit can be used to qualitatively normalize the distance between the two values of that measurement obtained in the hypothetical and observed "worlds". If the hypothetical and observed values are split by the value obtained in a working circuit, then the distance between the hypothetical and observed value is qualitatively large and therefore constitutes a counterexample to the given hypothesis.[12] If, however, the value obtained in the working circuit does not split the hypothetical and observed values, then the distance between them is a simple function (conditioned product) of (i) how far apart the two values are (their percentage difference) and (ii) the minimum of the differences between each of them and the working circuit value.

Assuming that a given measurement is not contradicted by the hypothesis, there is the issue of deciding if it actually supports the hypothesis or is just independent of it. This decision is reached by seeing if the value of that measurement in the correctly functioning circuit is qualitatively similar to values obtained in the hypothetical and observed "worlds". If it is, then the given measurement does not reflect any symptoms of either the actual fault or the hypothetical fault and is therefore independent of the hypothesis; but, if the value in the working circuit qualitatively differs from the other two similar values, then that measurement supports the hypothesis.

C. Hypothesis Generation (Theory Formation)

One of the more difficult logical tasks performed by SOPHIE is determining the set of possible faults or hypotheses that are consistent with the observed behavior of the faulted instrument (i.e., all the measurements taken up to that time). Such a capability is useful for several reasons. First, it can be called on by a student either when he has run out of ideas as to what could be wrong (i.e., what faults have not yet been ruled out by his measurements) or when he wishes to understand the full implications of his last measurement. It can also be called on by a tutorial specialist which might use this facility to detect the subtle ramifications of a measurement just performed by the student and thereby decide to query him

---

[12]This is the case unless the absolute difference between these measurements is less than some given threshold.

as to its significance. In a somewhat similar fashion, this facility already plays a major role in judging the quality of a given measurement (as will be discussed later) and in principle could be used to troubleshoot the instrument automatically.

The method of constructing the set of hypotheses uses the venerable "generate and test" paradigm: first, a backward working specialist, called the PROPOSER, examines the value observed for an external measurement and, from that observation, determines a list of all possible significant hypotheses which more or less explain that one measurement. This specialist uses a procedural form of simple production rules to encode its limited knowledge. Because the PROPOSER is not endowed with enough knowledge to capture all the complex interactions and subtleties of the circuit, it often errs by including a hypothesis that does not explain the observed behavior. In other words, the list it produces is overgeneral.

It is then the job of another specialist called the REFINER to take this overgeneral list and refine it. The REFINER, in essence, "simulates" each fault on the PROPOSERs list to make sure that it not only explains the output voltage (as a check on the PROPOSER) but also that it explains all the other measurements that the student has taken. By having the REFINER simulate each hypothesis, it takes into consideration all the complex interactions that a linear theory of the circuit fails to capture.

Counting on the simulator to check out all the subtle consequences of a proposed "theory" or hypothesis, however, leads to one major problem. For the REFINER to be able to simulate a hypothesis, the hypothesis must specify an explicit fault or modification to the circuit; but often the PROPOSER (like people) generates a *fault schema* which represents an infinite but structurally similar class of faults. For example, the hypothesis "the beta of the Darlington amplifier (of the IP-28) is low" is one such fault schema as is the hypothesis "C2 is leaky". For the first hypothesis it is not clear what the proposed beta is, just that it is lower than it should be, and for the second hypothesis, it is not specified how leaky C2 is or what its leakage resistance is supposed to be. In other words, a fault schema is an underspecified fault which has at least one unspecified parameter in its schema definition.

It is the job of another specialist, called the INSTANTIATOR, to take a fault schema and fill out or instantiate the unspecified parameters as best it can using an incomplete theory of the circuit. Once these parameters are instantiated, the fully specified hypothesis must then be checked to see if it really accounts for all the known measurements. For example, a subtle situation can arise where given any one context frame of measurements the proposed fault schema can be instantiated so as to be consistent with all the information derivable from that context frame. If, however, we simultaneously consider two or more context frames, we might discover there exists no consistent instantiation of the schema (i.e., the instantiated value created for one frame does not equal the value created for the second frame).

The INSTANTIATOR uses several techniques to determine a potentially consistent specification of a fault schema, the most general of which is a simple hill climbing strategy in which a specific value for the fault schema is guessed and then partially simulated (that is, the output voltage is determined). From the result of that simulation another guess is made until finally a value is found that causes the desired behavior in the given context frame.

As must be apparent, the hypothesis generator's numerous calls to the "simulator" could cause SOPHIE to consume countless cpu minutes before generating a set of viable theories. To avoid this, the REFINER and INSTANTIATOR use, in addition to the full-blown circuit simulator, a hierarchical, functional-block simulator. This latter simulator can either execute a functional block in the context of the whole circuit or simply in isolation as happens when the INSTANTIATOR wants to determine only the local effects of an instantiated fault schema. By correctly coordinating and maintaining consistency between these multiple representations of the circuit, several orders of magnitude of speed (over using just one simulator) can be realized.

Before leaving this section, it is worth noting that the INSTANTIATOR can be used in quite subtle ways to rule out certain faults even when no specific symptom has yet been encountered. For example, suppose the correct output voltage has been determined under a given load.   The INSTANTIATOR could use this fact to determine a *range of* possible values for a given fault schema such as effective beta of the Darlington amplifier (i.e., it can determine a lower bound of the combined beta such that if the beta had been any lower, the output voltage would have dropped and hence been symptomatic). Suppose then that the lower bound of this beta range was greater than betas of each of the two transistor making up the Darlington. This fact, in turn, would imply that neither of these transistors could be shorted! (Note then under many situations, one of these transistors could be shorted without there being any external symptoms.)

D. Determining Useless Measurements

Perhaps the most complex logical and tutorial tasks performed by SOPHIE and also one which illustrates a novel use of hypothesis generation is the task of verifying whether or not a given measurement *could* possibly add any new information to what is already known. If, for example, the result of a given measurement could be logically deduced from the previous measurements (using all the axioms of electronics, a complete description of the circuit, and a description of all the inputs and all possible faults, etc.), then this measurement could add no new information. In other words, this measurement would be logically redundant with respect to the prior measurements.

A moment's consideration reveals the potential complexity of proving that a measurement is redundant. Such a proof must take into consideration not only all the structural properties of each module making up the instrument but the functional relations between that module and all other modules as well. That is, it must take into account the global "purpose" of each component and each module in the overall design of the circuit!

Our approach to handling this task stems from an analogy to how one might prove the independence of an axiom set. If one is trying to establish that a new axiom is independent from a given collection of axioms, one might try rather to deny it by constructing a world model (or said differently, a "possible world") for the original axiom set which is not a world model for the augmented axiom set. Likewise one could consider the set of all "world models" or possible worlds for the original axiom set and determine if this set is "reduced" by adding the new axiom to the original set.

In our case, we view each measurement (along with its value) as an axiom or assertion and proceed according to the above analogy. The set of hypotheses constructed by the hypothesis generator serves as the set of all possible "world models" which satisfy or are consistent with the known measurements. Now a new measurement is taken and its value likewise passed to the hypothesis generator. If the resultant set of possible worlds is a proper subset of the prior set, then this measurement has added new information by eliminating at least one of the possible worlds (i.e., faults) and is therefore not redundant. Likewise if the set of possible worlds is the same (i.e., before and after the measurement was taken), this last measurement has added no information and is therefore logically redundant.[13]

E. Combining Qualitative and Quantitative Models of Knowledge

One of the more exciting possibilities for expanding SOPHIEs capabilities is in the interfacing of rule-based qualitative models of knowledge with SOPHIEs more quantitative models. Although there are many good reasons

---

[13]The above argument is, of course, not a proof and is only intended to be suggestive. In fact, since the hypothesis generator constructs only single fault possible "worlds" the above analogy is not literally true. Nevertheless, if the user is told that the circuit has only a single fault, then his space of possible worlds coincides with the hypothesis generator's space. The argument also becomes much more complex if the circuit has memory. In general, it must be shown that there cannot exist two measurements which taken individually do not rule out a possible world but taken collectively, do.

for investigating the interplay of qualitative and quantitative models of knowledge (deKleer, 1975), ours is driven by the awareness that our basic inferencing scheme has a major drawback. Namely, it achieves most of its answers without being able to generate a description of why the answers are true (i.e., a proof) or how the answer was derived. For example, SOPHIE can decide if a measurement is redundant, but it cannot "explain" or justify its decision in terms of causal reasoning. Indeed, part of the efficiency of our reasoning paradigm stems precisely from this fact. (See Chapter 4 for a theoretical discussion of this point.)

As a result we are beginning to investigate ways to combine an incomplete but qualitative (rule-oriented) type inference system with the "complete" model-driven schemes detailed in this chapter. In particular, we are intrigued with the possibility of using an "incomplete" qualitative theory to create a rationalization for an answer derived by the quantitative model-driven scheme. For example, once SOPHIEs hypothesis evaluator identifies which measurements contradict a given hypothesis, it seems like a much easier task to then "explain" why these measurements are counterexamples. Likewise, in handling hypothetical questions we do not need to count on the qualitative rules to sort out what does happen from what might plausibly happen. The quantitative models can do that.[14] Once we know for sure what happens, however, we can then use the qualitative rules for generating plausible causal explanations. Note that such an explanation can be useful even if it is not logically complete. It just has to highlight certain steps in the causal chain of reasoning.

Another possible direction in which to investigate combining quantitative and qualitative models of knowledge is to construct qualitative models for handling the ac and transient aspects of a circuit. In particular, the dc-based

---

[14]Note that actually we have here the chance to combine two completely different uses of qualitative reasoning. The first kind is used to create the interesting "examples" which are then passed off to the simulator. Once the simulator does its thing the second kind of qualitative reasoning could be called on to help explain the answer!

quantitative models can be used to determine the operating points of transistors. This information can then be used by qualitative ac specialists to predict such properties as clipping, distortion, etc. Likewise, a qualitative model could call a quantitative model to resolve any encountered ambiguities such as a feedback situation in which it is not clear, from a purely qualitative point of view, which of two opposing forces actually wins. In fact some of the original ideas for SOPHIE grew out of wanting to make such extensions to our purely qualitative reasoning scheme which used augmented finite state automata to model the qualitative properties and interactions of processes (Brown, Burton, & Zdybel, 1973).

IV. CONCLUSION

SOPHIE is sufficiently operational that it is ready for experimental use in a realistic instructional environment. Although it is a large and complex system, it is surprisingly fast, yielding response times in the order of a few seconds on a lightly loaded TENEX and requiring typically two cpu minutes per hour of use. We believe that much of this efficiency is due to: (i) the use of multiple representations of the constant or universal portions of SOPHIEs knowledge; (ii) the use of simulation as a general synthesis procedure for generating a world state description (or data base) which satisfies a given set of constraints; (iii) the close coupling of the structure of the world state description with the analysis or question-answering procedures that operate on it; and (iv) the use of heuristic strategies for expanding the domain of applicability of this highly tuned example-based reasoning paradigm so that it can handle the complex tasks of hypothesis evaluation, hypothesis generation, and redundancy checking. The ability of this scheme to handle complex, multistate feedback systems is encouraging since it is precisely these kinds of "worlds" that are most difficult to capture within the classical AI paradigms.

ACKNOWLEDGMENTS

We are indebted to Alan G. Bell for his extensive help in building the simulators and the hypothesis generation system and to Rusty Bobrow for the countless hours he has spent discussing the theoretical ramifications of SOPHIE. Much of our current view of SOPHIE stems from his observations. This work was supported in part by the Advanced Technical Training program ARPA-HRRO, and Air Force Human Resources Laboratory Technical Training.

REFERENCES

Brown, J. S., Burton, R. R., & Bell, A. G. SOPHIE: A step towards a reactive learning environment. *International Journal of Man-Machine Studies*, in press, 1975.

Brown, J. S., Burton, R. R., & Bell, A. G. SOPHIE: A sophisticated instructional environment for teaching electronic troubleshooting (an example of AI in CAI). (BBN Report No. 2790). Cambridge, Mass.: Bolt Beranek and Newman, Inc., March, 1974.

Brown, J. S., Burton, R. R., & Zdybel, F. A model-driven question answering system for mixed-initiative computer assisted instruction. *IEEE Transactions on Systems, Man and Cybernetics*, May, 1973, *SMC-3*.

Burton, R. R. Semantically centered parsing. Ph.D. Dissertation, University of California at Irvine, December, 1975.

Carbonell, J. R. AI in CAI: an artificial intelligence approach to computer-assisted instruction. *IEEE Transactions on Man-Machine Systems*, December, 1970, *MMS-11* 190-202.

deKleer, J. Qualitative and quantitative knowledge in classical mechanics. Working Paper, MIT-AI Laboratory, February, 1975.

Nagel, L. W., & Pederson, D. O. SPICE: simulation program with integrated circuit emphasis. Memorandum ERL-M382, Electronics Research Laboratory, University of California at Berkeley, April, 1973.

Teitelman, W. INTERLISP reference manual. Xerox Palo Alto Research Center, October, 1974.

Winograd, T. *Understanding natural language*. New York: Academic Press, 1972.

Woods, W. A. Transition network grammars for natural language analysis. *Communications of the Association for Computing Machinery*, 1970, *13*(10), 591-606.

THE ROLE OF SEMANTICS
IN AUTOMATIC SPEECH UNDERSTANDING

Bonnie Nash-Webber
Bolt Beranek and Newman
Cambridge, Massachusetts

I. INTRODUCTION

In recent years, there has been a great increase in research into automatic speech understanding, the purpose of which is to get a computer to understand the spoken language. In most of this recent activity it has been assumed that one needs to provide the computer with a knowledge of the language (its syntax and semantics) and the way it is used (pragmatics). It will then be able to make use of the constraints and expectations which this knowledge provides, to make sense of the inherently vague, sloppy, and imprecise acoustic signal that is human speech.

Syntactic constraints and expectations are based on the

patterns formed by a given set of linguistic objects, e.g., nouns, verbs, adjectives, etc. Pragmatic ones arise from notions of conversational structure and the types of linguistic behavior appropriate to a given situation. The bases for semantic constraints and expectations are an *a priori* sense of what can be meaningful and the ways in which meaningful concepts can be realized in actual language.

We will attempt to explore two major areas in this chapter. First we will discuss which of those things that have been labeled "semantics" seem necessary to understanding speech. We will then expand the generalities of the first section with a detailed discussion of some specific problems which have arisen in our own attempts at understanding speech and some semantics-based mechanisms for dealing with them.

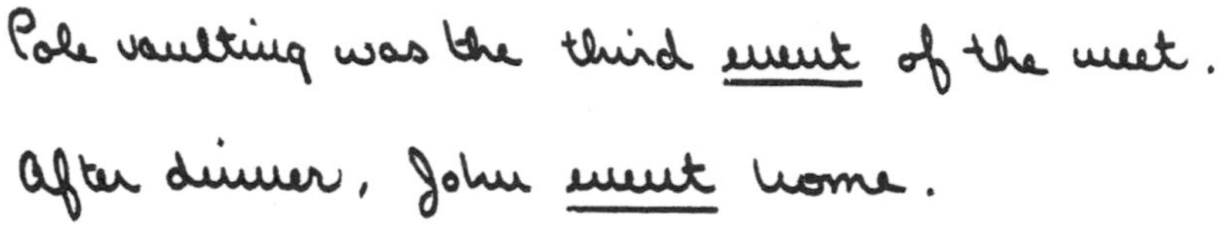

Pole vaulting was the third event of the meet.

After dinner, John went home.

Fig. 1. Ambiguous handwriting sample.

Psychologists have demonstrated that it is necessary for people to be able to draw on higher-level linguistic and world knowledge in their understanding of speech: the acoustic signal they hear is so imprecise and ambiguous that even a knowledge of the vocabulary is insufficient to insure correct understanding. For example, Pollack and Pickett's (1964) experiments with fragments of speech excised from eight-word sentences and played to an audience showed that 90% intelligibility was not achieved until a fragment spanned six of the eight words, and its syntactic and semantic structures were becoming apparent. [See Wanner (1973) for an excellent survey of psycholinguistic experiments on listening and comprehension.] When the acoustic signal is ambiguous out of context, as a signal equally interpretable as "his wheat germ and honey" or "his sweet German honey", higher-level knowledge must be used to force one reading over the other. A similar problem arises in understanding handwritten text. Notice in Fig. 1 how the same scrawl is recognized as two different words in contexts engendering different predictions. Without making any claims about how a person actually understands speech, this chapter will discuss some aspects of semantic knowledge and how they seem to contribute to understanding speech by a computer. We then discuss specific problems in our attempts at understanding speech, and some semantics-based mechanisms for dealing with them.

II. FORMS OF SEMANTIC KNOWLEDGE

If a speech understander must use semantic knowledge to constrain the ways of hearing an utterance, then his knowledge must represent what can be meaningful and expected at any point in a dialog. Let us consider what types of semantic knowledge determine what is meaningful and enable predictions.

A. Knowledge of Names and Name Formation

Knowledge of the familiar names and models for forming new compound names permits a listener to expect and hear meaningful phrases. For example, knowing the words "iron" and "oxide" and what they denote, and that a particular oxide (or set of them) may be specified by modifying the word "oxide" with the name of a metal may enable a listener to hear the sequence "iron oxides", rather than "iron ox hides", "I earn oxides", or even "Ira knocks sides".

B. Knowledge of Lexical Semantics

Knowledge of lexical semantics (models of how words can be used and the correspondence between concepts in memory and their surface realizations) enables the listener to predict and verify the possible surface contexts of particular words. Given a hypothesis that a word has occurred in the utterance, a local prediction can be made of what words could appear to its left or right. For example, when the word "contain" appears in a sentence (or any other lexical realization of the concept CONTAINMENT), two other concepts strongly associated with it - a container and a containee - should also do so. These might be called the "arguments" to CONTAINMENT. When "contain" is used as the main verb of an active sentence, its subject must be a location or container, and its object must refer to something which can be contained. When the concept of CONTAINMENT is realized within a noun phrase, these arguments can appear in several ways. For example, in "the silicon content of sample 3", the container follows in a prepositional phrase (with the preposition "in" or "of") and the object being contained precedes as a noun classifier; in "sample 3's silicon content", the container also precedes, this time as a possessive.

While we can, however, profitably use lexical semantics to predict the local context of a word from the concepts it can partially instantiate, enumerating all possible word sequence realizations of a concept rarely pays. There are usually too many possibilities available. For example, the concept of CONTAINMENT is realized in all the following phrases:

> rocks containing sodium,
> sodium-rich basalts,
> igneous samples with sodium,
> samples in which there is sodium,
> rocks which have sodium.

C. Knowledge of Conceptual Semantics

Knowledge of conceptual semantics, how concepts are associated in memory, facilitates "global" predictions between utterances, as well as local ones with a single utterance. Some global predictions relate to a "natural" flow of conversation: if one concept is under discussion,

certain other ones will likely appear. Other predictions
involve expectations of which related concepts can be left
implicit in the context, and not verbalized in the discourse;
these predictions help in accommodating ellipsis and
anaphora. A short example of conversation should suffice
here to illustrate the point:

> "I'm flying to New York tomorrow. Do
> you know the fare?"
> "About 26 dollars each way."
> "Do I have to make reservations?"
> "No."
> "Super."

First, the concept of a trip is strongly linked with such
other concepts such as destinations, fares, transportation
mode, departure date, etc., so one might expect them to be
mentioned in the course of a conversation about a trip.
Secondly, the strength of these associations is both domain,
context, and user dependent. If the domain concerns
planning trips, as in making airline reservations, then
destination and departure date would seem to have the
strongest links with trips. In another domain such as
managing the travel budget for a company, it may only be
the cost of the trip and who is paying for it that have this
strong association. As far as context and user dependency,
the company accountant's primary interest in business trips
may be quite different from that of a project leader
wondering where his people are going.

The places where ellipsis is most likely to occur seem to
correlate well with strong interconcept associations. This is
useful information since it suggests when *not* to look hard
for related concepts in the local context. For example, "the
fare" and "reservations" are both elliptical phrases: "the
fare" must be for some trip via some vehicle at some time,
but fare is so strongly linked to a flight that is is not
necessary to mention a flight locally if it can be
understood from context, as in, "Do you know the *current
air* fare *to New York*?" Knowledge of the concepts associated
with trips and fares and how "strong" the links are, are
needed to make these local or global predictions.

All global expectations are not semantics-based, of

course. Rhetorical devices available to a speaker, such as parallelism and contrast, add to these global expectations about future utterances. In addition, problem solving situations also have a strong influence on the nature of discourse and the speaker's overall linguistic behavior.

D. Knowledge of the Use of Syntactic Structures

Knowledge of the meaningful relations and concepts which different syntactic structures can convey enables the listener to rescue cues to syntactic structure which might otherwise be lost. Among the meaningful relations between two concepts, A and B, that can be communicated syntactically are that B is the location of A, the possessor of A, the agent of A, etc. Other syntactically communicated concepts include set restriction (via relative clauses), eventhood (via gerund constructions), facthood (via "that"-complements), etc. Syntactic structure is often indicated by small function words (e.g., prepositions and determiners) which have very imprecise acoustic realizations. The knowledge of what semantic relations can meaningfully hold between two concepts in an utterance and how these relations can be realized syntactically can often help in recovering weak syntactic cues. On the other hand, one's failure to recover some hypothesized cue, once attempted, may throw doubt on one's semantic hypothesis about the utterance.

E. Knowledge of Specific Facts and Events

Knowledge of specific facts and events can also be brought in as an aid to speech understanding, though it is less reliable than the other types of semantic knowledge discussed above. This is because it is more likely that two people share the same sense of what is meaningful than that they are aware of the same facts and events. Fact and event knowledge can be of value in confirming, though not in rejecting, one's hypotheses about an utterance. For example, if one knows about Dick's recent trip to Rhode Island for the Americas Cup, and one hears an utterance concerning some visit Dick had made to -- Newport?, New Paltz?, Norfolk?, Newark? -- one would probably hear, or chose to hear, the first, knowing that Dick had indeed been to Newport. One could not, however, reject any of the others, on the grounds that the speaker may have more information than the listener.

F. Knowledge of Errors

The kinds of errors that frequently occur in speech must be accounted for in any valid model of human language understanding. The errors occur at all linguistic levels -- phonemic, syntactic, and semantic. Ones seemingly related to semantic organization (because the meaning of the resulting utterance seems close to the supposed intention of the speaker) include malapropisms, portmanteaus, mixed metaphors and idioms, etc. For example, "I'm glad you reminded me: it usually takes a week for something to sink to the top of my stack." ["sink in" - "rise to the top of the stack"]; or "Follow your hypothesis to its logical confusion." ["logical conclusion"] [See Fromkin (1971) for additional examples.] These errors rarely occur in text, whose production is much more deliberate and considered than that of speech.

Given that we have decided that our reading of part or all of an utterance must be wrong, we must be able to suggest where the source of the error lies and what the best alternative hypothesis is. Moreover we must do so efficiently, lest we fail to come up with a satisfactory reading in reasonable time. The works of Marcus (1974), Riesbeck (1974), Sussman (1973), Winograd (1969) and Woods (1970) present several different schemes for dealing with these problems of error analysis and correction.

III. RECENT SPEECH UNDERSTANDING RESEARCH

In those initial attempts at automatic speech understanding which use some sort of semantic knowledge [SDC's VDMS - (Barnett, 1973; Ritea, 1973); CMU's HEARSAY I - (Neely, 1973; Reddy, Erman, Fennell, &

Neely, 1973); Lincoln Lab's VASSAL - (Forgie, 1973); SRI's
system - (Walker, 1973a,b); BBN's SPEECHLIS - (Woods,
1974; Rovner, Nash-Webber, & Woods, 1973)], one sees more
similarities than dissimilarities in their view of semantic
knowledge.   All aspire to use knowledge of constraints on
acceptable and appropriate utterances to work outwards
from what they are relatively sure of to what is initially
more vague and ambiguous.   Some of the systems work left-
to-right in this process, while others work in both
directions from whatever strong "anchor points" they can
identify. Most of the systems also make predictions about
the likely content of an incoming utterance before
analyzing the utterance itself. This intuitively seems to
reflect human speech understanding. Unfortunately, none
of these initial systems seems to do it very well.   In
addition, all of the systems have chosen one or two small
interactive task domains about which the user is expected
to speak and the system is counted upon to respond
appropriately. This enables them to take greater advantage
of all the constraints on meaningful utterances discussed
earlier.
    The main differences lie in the amount of semantic
knowledge they attempt to represent and the form in which
that knowledge is represented. For instance, since the
degree to which natural language syntax has been captured
in various formalisms far exceeds that for semantics, several
of these systems have captured their semantics in terms of
syntax in a categorial grammar (one whose elements are
semantic in character like "rockname" or "color", rather
than syntactic like "noun" or "adjective"). Current research
in semantic formalisms and representations of knowledge
should result in more sophisticated semantics being more
widely used in systems for understanding speech.

IV. THE SPEECHLIS ENVIRONMENT

    We shall now give a detailed discussion of SPEECHLIS,
the BBN speech understanding system to point out many
interesting specific problems in speech understanding. The
deficiencies we note in our current solutions may help the
reader avoid having to discover these deficiencies for

himself, and may suggest to him better solutions. In addition, the framework presented may be suggestive to psychologists and psycholinguists attempting to discover and explain the mechanisms by which humans understand language.

A. Task Domain

Currently, there are two domains in which the problems of speech understanding are being studied: a natural language information retrieval system for lunar geology and a similar one for travel budget management. Typical requests in the two domains include:

Lunar Geology
> Give the average K/Rb ratio for each of the fine-grained rocks.
> Which samples contain more than 10ppm aluminum?
> Has lanthanum been found in the lunar fines?

Travel Budget Management
> How much was spent on trips to California in 1973?
> Who's going to the ACL meeting in June?
> If we only send two people to IFIP-74, will we be within our budget?

It is envisioned that a user will carry on a spoken dialog with the system in one of these areas in order to solve some problem. The reason for choosing the former area was to draw on a two-year experience with the BBN LUNAR system (Woods, Kaplan, & Nash-Webber, 1972). The latter permits investigation of the problems of user and task modeling, which turn out to be very inconvenient in the specialized technical area of lunar geology. The lunar geology vocabulary for the SPEECHLIS system contains about 250 words, of which approximately 75 are "function words" (determiners, prepositions, auxiliaries and conjunctions) and the remaining 175 are semantically meaningful "content words". The vocabulary for travel budget management is larger, containing about 350 words, with approximately the same core of 75 function words.

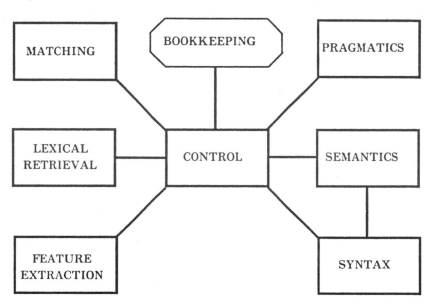

Fig. 2. Components of a speech-understanding system.

B. System Organization

The basic structure of SPEECHLIS is given in Fig. 2, with its components pictured as labeled boxes and the control and communication paths represented as lines between them. Because the inherent ambiguity and vagueness of the speech waveform do not permit its unique characterization by the acoustic-phonetic component, it is possible for many words to match, to some degree, any region of the input. In order to record these possibilities, we have the lexical retrieval and word matching components produce a *word lattice*, whose entries are words which were found to match well (i.e., above some threshold) in some region of the utterance. Associated with each word match is a description of how it matched the input and the degree of confidence in the match. (A sample word lattice is shown in Fig. 3. The horizontal axis represents time; the integer labeled boundaries divide the utterance into segments which can possibly be characterized as a single speech sound or phoneme. Though they look equal in the figure, these segments need not span the same length of time.) Since small words like "an", "and", "in", "on", etc. tend to lose their phonemic identity in speech and result in spurious matches everywhere, we avoid these spurious matches by initially trying to match only words of three or more phonemes in length. The motivation for looking at all strong matches at once, rather than in a strict left-to-right order, is that there are too many syntactic and semantic possibilities if we cannot use the really good matches of content words to suggest what the utterance might be about.

This initial, usually large lattice of good big word matches then serves as input to the syntactic, semantic, and pragmatic components of the system. Subsequent processing involves these components working, step by step, both separately and together, to produce a meaningful and contextually apt reconstruction of the utterance, which is hoped to be equivalent to the original one. Steps in proposing or choosing a word reflect some hypothesis about what the original utterance might be. In SPEECHLIS, this notion of a current hypothesis is embedded in an object we call a *theory*, which is specifically a hypothesis that some set of word matches from the word lattice is a partial (or complete) reconstruction of the utterance. Each step in the higher-level processing of the input then involves the creation, evaluation, or modification of such a theory.

The word lattice is not confined, however, to the initial set of "good, long" word matches. During the course of processing, any one of the higher-level components may make a *proposal*, asking that a particular word or set of words be matched against some region of the input, usually adjacent to some word match hypothesized to have been in the utterance. The minimum acceptable match quality in this case would be less than in the undirected matching above for two reasons. First, there would be independent justification from the syntax, semantics, and/or pragmatics components for the word to be there, and second, the word

362

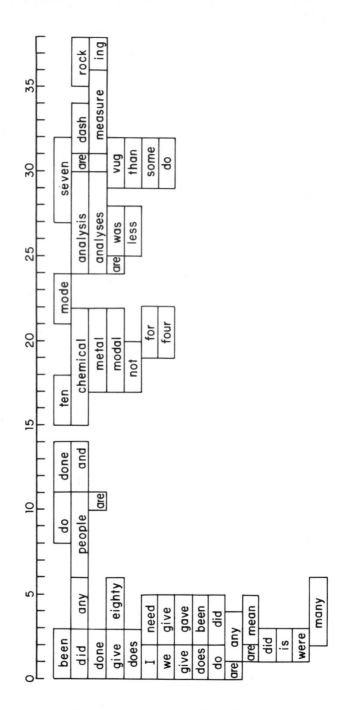

Fig 3. A sample word lattice.

may have been pronounced carelessly because that independent justification for its existence was so strong. For example, take a phrase like "light bulb", in ordinary household conversation. The word "light" is so strongly predicted by bulb in this environment, that its pronunciation may be reduced to a mere blip that something preceded "bulb". In the case of proposals made adjacent to, and because of, some specific word match, the additional information provided by the phonetic context of the other word match will usually result in a much different score than when the proposed word is matched there independent of context.

The control component (see Fig. 2) governs the formation, evaluation and refinement of theories, essentially deciding who does what when, while keeping track of what has already been done and what is left to do. It can also take specific requests from one part of the system that another part be activated on some specific job, but retains the option of when to act on each request. (In running SPEECHLIS with early versions of the control, syntactic and semantic components, we found several places where for efficiency, it was valuable for Syntax to be able to communicate directly with Semantics during parsing, without giving up control. (N.B. We will be using initial capitals on the words "syntax", "semantics" and "pragmatics" when referring to parts of SPEECHLIS.) Thus, it is currently also possible for Syntax to make a limited number of kinds of calls directly to Semantics. How much more the initial control structure will be violated for efficiency's sake in the future is not now clear.)

The reason that processing does not stop after initial hypotheses have been formed about the utterance is that various *events* may happen during the analysis of a theory which would tend to change SPEECHLIS's confidence in it, or cause SPEECHLIS to want to refine or modify it. For example, consider some utterance extracted from a discussion of the lunar rocks. Under the hypothesis that the word "lunar" occurred in the utterance, a good match found for "sample" to its right would greatly increase our confidence that both words were actually there in the original utterance. An entity called an *Event Monitor* can be set up as an active agent during the processing of a theory by some higher-level component, to watch for some particular event which would change that component's opinion of the theory. When such an event occurs, the monitor will create an appropriate *Notice*, describing its particular characteristics and its approximate worth. Notices are sent to the control component which decides if and when to act on them. Only when a notice is acted upon will the appropriate revaluation, refinement or modification occur. Examples of semantic monitors and events will be found further on in this chapter.

To summarize then, the semantics component of SPEECHLIS has available to it the following facilities from the rest of the system: access to the words which have been found to match some region of the acoustic input and information as to how close to the description of the input that match is; ability to ask for a word to be matched against some region of the input; and ability to build or flesh out theories based on its own knowledge and to study those parts of a theory built by Syntax and Pragmatics. Given this interface with the rest of the SPEECHLIS world, how does Semantics make its contribution to speech understanding?

V. SPEECHLIS SEMANTICS

The primary source of permanent semantic knowledge in SPEECHLIS is a network of nodes representing words, "multiword names", concepts, specific facts, and types of syntactic structures. Hanging onto each concept node is a data structure containing further information about its relations with the words and other concepts it is linked to, and which is also used in making predictions. The following sections describe the network nodes and the ways in which semantic predictions are enabled.

A. Network-Based Predictions

Multiword Names. Each content word in the vocabulary (i.e., words other than articles, conjunctions, and prepositions; for example, "ferric", "iron", "contain") is associated with a single node in the semantic network. It is linked to phonetic representations of the word. A word node is also linked to nodes representing "multiword names" of which the original word is a part. For example, "fayalitic olivine" is an multiword name linked to both "fayalitic" and "olivine"; "fine-grained igneous rock" is one linked to the word "fine-grained" and the multiword name "igneous rock". This allows the phonetic representations to be shared.

Representing multiword names in this way has two advantages. First it enables us to maintain a reasonable size phonemic dictionary in SPEECHLIS (i.e., by not having to store the pronunciations of compounds like "fayalitic-olivine" and "principal-investigator" as well as the pronunciations of the individual words). Second, it enables us to make local predictions. That is, any given word match may be partial evidence for a multiword name of which it is a part. The remaining words may be in the word lattice, adjacent and in the right order, or missing due to poor match quality. In the former case, one would eventually notice the adjacency and hypothesize that the entire multiword name occurred in the original utterance. In the latter case, one would propose the missing words in the appropriate region of the word lattice, with a minimum acceptable match quality directly proportional to the urgency of the success of the match. That, in turn, depends on how necessary it is for the word in the match to be part of a multiword name. That is, given a word match for "oxide", Semantics would propose "ferrous" or "ferric" to its left, naming "ferrous oxide" or "ferric oxide". Given a match for "ferric" or "ferrous", Semantics would make a more urgent proposal for "oxide", since neither word could appear in an utterance alone. Further details on the proposing and hypothesizing processes will be given in the following sections.

There is another advantage to representing multiword names in this way rather than as compound entries in the dictionary. As an immediate consequence, it turns out that fayalitic olivine is a type of olivine, a fine-grained igneous rock is a type of igneous rock which is a type of rock, and a principal investigator is a type of investigator. No additional links are needed to represent this class information for them.

One might argue, of course, that this implicit class association is not always appropriate. For example, "fire dogs", "sun dogs", and "hot dogs" are not dogs in any sense. For most compounds, however, an acquired idiomatic sense does not completely replace the original one, which is most often associational. Else why form the compound name in that way? For efficiency then, because phonetic and associative cues usually agree, we choose to take advantage of the implied class association. When it is not appropriate, the compound is made a separate lexical entry.

Concept-Argument Relations. From the point of view of Semantics, an action or an event is a complex entity, tying several concepts together into one that represents the action or event itself. Syntactically, an action or event can be described in a single clause or noun phrase, each concept realizing some syntactic role in the clause or phrase. One of these concepts is that associated with the verb or nominal (i.e., nominalized verb) which names the *relation* involved in the action or event. The other concepts serve as *arguments* to the relation. For a verb, this means they serve as its subject, object, etc.; for a nominal, it means they serve as premodifiers (e.g., adjectives, noun-noun modifiers, etc) or as postmodifiers (e.g., prepositional phrases, adverbials, etc.) For example,

John went to Santa Barbara in May.
SUBJ VERB PREP-PHRASE PREP-PHRASE

John's trip to Santa Barbara in May.
PREMOD NOMINAL PREP-PHRASE PREP-PHRASE

In the semantic network, an action or event concept is linked to the one which names the relation and the ones which can fill its arguments.

Semantics uses its knowledge of words, multiword names, and concepts to make hypotheses about possible local contexts for one or more word matches, detailing how the word matches fit into that context. Given a word match, Semantics follows those links in the network which lead from the word to concepts it is an instance of and also to multiword names and concepts which it may partially instantiate. On each of the nodes which represent other components of the partially instantiated name or concept, Semantics sets an *event monitor*. In following network links for another word match, should a monitored node be instantiated (and conditions on the instantiation specified in the monitor be met), an *event notice* is created, calling for the construction of a new, expanded theory.

To see this, consider the network shown in Fig. 4 and a word match for "oxide". Since "oxide" occurs in the multiword names "ferrous oxide" and "ferric oxide", Semantics would set event monitors on the nodes for "ferrous" and "ferric", watching for either's instantiation to the immediate left of "oxide". It would also propose them there. The net also shows that oxides can be constituents of rocks and a rock constituent can serve as one argument to the concept CONTAIN, specifically the thing being contained. (The other argument, the container, is realizable by the concept SAMPLE.) Semantics would therefore set a monitor on the node for CONTAIN and one on the node for SAMPLE. If Semantics is later given a word match for "contain" or one of its inflected forms, or one which instantiates SAMPLE (e.g., "rock"), it would be seen by the appropriate monitor when it reaches the node for CONTAIN (or SAMPLE), and result in the creation of an event notice linking "oxide" with the new word match.

Each notice contains a weight, currently an integer, representing both how confident Semantics is that the resulting theory is a correct hypothesis about the original utterance and how useful to Semantics that theory would be. For example, Semantics would assign a higher weight to an event notice linking "contain" and "oxide" than to one linking "rock" and "oxide" for two reasons. First, Semantics is less certain that a theory for "rock" and "oxide" will eventually instantiate the concept CONTAIN than it is that a theory for "contain" and "oxide" will do so. That is, because there are many other possible ways of instantiating both SAMPLE and CONSTITUENT, but only "contain" or one of its inflections can instantiate the head of CONTAIN. Secondly, there are many different concepts which "rock" and "oxide" can together partially instantiate, but only one which "contain" and "oxide" can -- that of CONTAINMENT. While further semantic processing on a theory for "contain" and "oxide" would only involve seeking an instantiation of SAMPLE, further processing of one for "rock" and "oxide" would involve trying many possibilities. Currently this weight is only a rough measure of confidence and usefulness, though it should become more sophisticated as we develop a better sense for what kinds of theories lead most efficiently to a complete understanding of the utterance.

Syntactic Structures. Nodes corresponding to the syntactic structures produced by the grammar (e.g., noun phrases, to-complements, relative clauses, etc.) are also used in making local predictions. First, if an argument to some concept can be specified as a particular syntactic structure with a particular set of syntactic features, we want to predict an occurrence of that structure, given an instantiation of the concept's head. For example, a concept headed by "anticipate" may have as its object an embedded sentence whose tense is future to the tense of "anticipate". We want to be able to predict and monitor for any such structures and notice them if built.

> I anticipate that we will make 5 trips to L.A.
> I anticipated that we would have made 5 trips to L.A. by November.

More generally, we want to be able to use any co-occurrence restrictions on lexical items and syntactic structures or features in making predictions. For example, when different time and frequency adverbials may be used depends on the mood, tense, and aspect of the main clause and certain features of the main verb. "Already", for instance, prefers that clauses in which it occurs, headed by a non-stative verb, be either perfective or progressive or both, unless a habitual sense is being expressed. For example,

> John has already eaten 15 oysters.
> John is already sitting down.
> ?John already ate 15 oysters.
>   (Perfective is preferable.)
> *John already sits down.
> John already runs 5 miles a day. (Habitual)

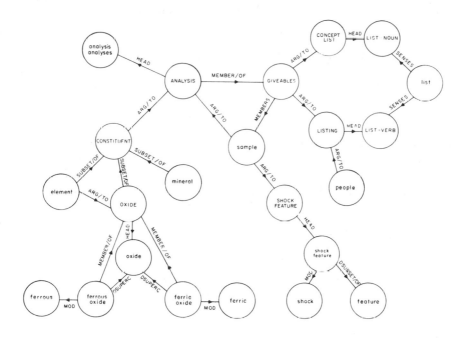

Fig 4. Small semantic network.

Secondly, if a concept with an animate agent as one of its arguments is partially instantiated, Semantics might want to predict some expression of the agent's purpose in the action. Now it is often possible to recognize "purpose" on syntactic grounds alone, as an infinitive clause introduced by "in order to", "in order for X to", "to", or "for X to". For example,

> John's going to Stockholm to visit Fant's lab.
> I need $1000 to visit Tbilisi next summer.
> John will stay home in order for Rich to finish his paper.

These syntactic structure nodes then facilitate the search for a "purpose": they permit monitors to be set on the semantic concept of PURPOSE, which can look for, *inter alia*, the infinitive clauses popped by Syntax.

B. Case Frame-Based Predictions

Description of a Case Frame. Additional information about how an action or event concept made up of a relation and its arguments may appear in an utterance is given in a *case frame*, a la Fillmore (1968), associated with the concept. Case frames are useful both in making local predictions and in checking that some possible syntactic organization of the word matches in a theory supports Semantics' hypotheses. Fig. 5 shows the case frames for the concepts ANALYSIS and CONTAIN.

Case frame for ANALYSIS

```
(((REALIZES . NOUN-PHRASE))
 (NP-HEAD (EQU . 14) NIL OBL)
 (NP-OBJ (MEM . 1) (OF FOR) ELLIP)
 (NP-LOC (MEM . 7) (IN FOR OF ON) ELLIP))
```

Case frame for CONTAIN

```
(((REALIZES . CLAUSE)
  (ACTIVSUBJ S-LOC)
  (PASSIVSUBJ S-PAT))
 (S-HEAD (EQU . 20) NIL OBL)
 (S-LOC (MEM . 7) (IN) OBL)
 (S-PAT (MEM . 1) NIL OBL))
```

```
Concept 14  -  concept of ANALYSIS
Concept  1  -  concept of COMPONENT
Concept  7  -  concept of SAMPLE
Concept 20  -  concept of CONTAIN
```

Fig. 5. Case frames for Analysis and Contain.

A case frame is divided into two parts: the first part contains information relating to the case frame as a whole;

the second, descriptive information about the cases. (In the literature, cases have been associated only with the arguments to a relation. We have extended the notion to include the relation itself as a case, specifically the head case (NP-HEAD or S-HEAD). This allows a place for the relation's instantiation in an utterance, as well as the instantiations of each of the arguments.)

Among the types of information in the first part of the case frame is a specification of whether a surface realization of the case frame will be parsed as a clause or as a noun phrase, indicated in our notation as (REALIZES . CLAUSE) or (REALIZES . NOUN-PHRASE). If as a clause, further information specifies which cases are possible active clause subjects (ACTIVSUBJ's) and which are possible passive clause subjects (PASSIVSUBJ's). In the case of CONTAIN (Fig. 5b), the only possible active subject is its location case (S-LOC), and the only possible passive subject is its patient case (S-PAT). For example,

Does each breccia contain olivine?
S-LOC   S-PAT

Is olivine contained in each breccia?
S-PAT   S-LOC

(While not usual, there are verbs like "break" which take several cases (here - agent, object, and instrument), and can appear with various of them as its active subject.

John broke the vase with a rock. (agent)
A rock broke the vase. (instrument)
The vase broke. (object)

The case which becomes the active subject is, however, rigidly determined by the set of cases actually present. In ACTIVSUBJ, the cases are ordered, so that the first one in the list which will occur in an active sentence will do so as the subject. For some reason, there appears to be no syntactic preference in selecting which case becomes passive subject, so the case names on PASSIVSUBJ are not ordered.)

The first part of the case frame may also contain such

information as intercase restrictions, as would apply between instantiations of the arguments to RATIO (i.e., that they be measurable in the same units).

The second part of the case frame contains descriptive information about each case in the frame:

a) its name, e.g., NP-OBJ, S-HEAD (The first part of the names gives redundant information about the frame's syntactic realization: "NP" for noun phrase and "S" for clause. The second part is an abbreviated Fillmore-type (Fillmore, 1968) case name: "OBJ" for object, "AGT" for agent, "LOC" for location, etc.);

b) the way it can be filled - whether by a word or phrase naming the concept (EQU) or an instantiation of it (MEM), e.g., (EQU . SAMPLE) would permit "sample" or "lunar sample" to fill the case, but not "breccia". Breccia, by referring to a subset of the samples, only instantiates SAMPLE but does not name it;

c) a list of prepositions which could signal the case when it is realized as a prepositional phrase (PP). If the case were only realizable as a premodifier in a noun phrase or the subject or unmarked object of a clause, this entry would be NIL;

d) an indication of whether the case must be explicitly specified (OBL), whether it is optional and unnecessary (OPT), or whether, when absent, it must be derivable from context (ELLIP). For example, in "The bullet hit.", the object case - what was hit - must be derivable from context in order for the sentence to be "felicitous" or well posed. (We plan to replace this static, three-valued indication of sentence level binding with functions to compute the binding value. These functions will try to take into account such discourse level considerations as who is talking, how he talks and what aspects of the concept he is interested in, and other factors such as case interaction. For example, the arguments to RATIO will be either both explicitly specified or both ellipsed, but only rarely will they go separate ways.)

Uses of Case Frames. Semantics uses case frame information for making local predictions in the form of proposals and checking the consistency of syntactic and semantic hypotheses. These predictions mainly concern the occurrence of a preposition at some point in the utterance or a case realization's position in an utterance relative to cases already realized. The *strength* of such a prediction depends on its cost: the fewer the words or phrases which could realize the case, and the narrower the region of the utterance in which to look for one, the cheaper the cost of seeking a realization. For example, since there are fewer words and phrases which name a concept (EQU marker) as opposed to instantiating it (MEM marker), cases marked EQU would engender stronger predictions. The *urgency* of the proposal depends on its likelihood of success, given the hypothesis is true: if the case must be realized in the utterance (OBL marker), the prediction should be successful if the initial hypothesis about the concept associated with the case frame is correct. If the case need not be present in the utterance (ELLIP or OPT marker), even if the initial hypothesis is correct, the prediction need not be successful. Both the strength and urgency of a proposal are taken into account by Control in deciding whether or not to have the Word Matcher attempt the suggested match.

Consider the case frame for ANALYSIS in Fig. 5a for example. If we believed that the word "analysis" occurred in the utterance, we would predict the following: (1) an instantiation of either COMPONENT or SAMPLE to its immediate left (that is, as a premodifier), (2) either "of" or "for" to its immediate right, followed by an instantiation of COMPONENT, and (3) either "in", "for", "of", or "on" to its immediate right, followed by an instantiation of SAMPLE. It does not matter that the above predictions are contradictory: if more than one were successful (i.e., there were more than one way of reading that area of the speech signal), it would simply be the case that more than one new theory would be created as refinements of the original one for "analysis", each incorporating a different alternative.

It is important to remember that in most cases we are predicting likely locations for case realizations, not necessary ones. If they fail to appear in the places predicted, it does not cast doubts on a theory. English allows considerable phrase juggling -- e.g., preposing prepositional phrases, fronting questioned phrases, etc., and, of course, not all predicted pre- and postmodifiers of a noun can occur to its immediate left or right. This must be remembered in considering how these local, frame-based predictions should be understood. The point is just that our confidence is greater when a case is realized in a likely place than in an unlikely one.

```
[ CFT#6
        (((REALIZES . NOUN-PHRASE))
        (NP-HEAD (ANALYSES . 14) NIL OBL)
        (NP-GOAL (CFT#5 . 1) (OF FOR) ELLIP)
        (NP-LOC (MEM . 7) (IN FOR OF ON) ELLIP))]


[ CFT#5
        (((REALIZES . NOUN-PHRASE)
        (CASEOF CFT#6))
        (NP-MOD (FERROUS . 13) NIL OBL)
        (NP-HEAD (OXIDE . 5) NIL OBL))]
```

Fig 6. Case frame tokens.

For example, consider Semantics processing a theory that a word match for "contain" was part of the original utterance. As mentioned earlier, "contain" heads the concept CONTAIN, whose other arguments are SAMPLE and CONSTITUENT. On both of these, monitors would be set to notice later instantiations of these concepts. Under the hypothesis that the clause is active, Semantics would include in the monitor set on the concept SAMPLE, the only possible active subject, that its instantiation be to the left of the match for "contain". In the monitor set on COMPONENT, the active object, we would indicate a preference for finding its instantiation to the right. This latter is only a preference because by question fronting, the object may turn up to the left, e.g., "What rare earth elements does each sample contain?". (Notice that regardless of where an instantiation of either SAMPLE or COMPONENT is found in the utterance, it will be noticed by the appropriate monitor. It is only how valuable the particular concept instantiation is to the theory setting the monitor that is affected by a positional preference.)

The process of checking the consistency of Syntax's and Semantics' hypotheses uses much the same information as that of making frame-based local predictions. As word matches are included in a theory, Semantics represents its hypotheses about their semantic structure in *case frame tokens*. These are instances of case frames which have been modified to show which word match or which other case frame token fills each instantiated case.

The two case frame tokens in Fig. 6 represent semantic hypotheses about how the word matches for "analyses", "ferrous" and "oxide" fit together. "Analyses" is the head (NP-HEAD) of a case frame token whose object case (NP-OBJ) is filled by another case frame token representing "ferrous oxide". Another way of showing this is in the tree format of Fig. 7.

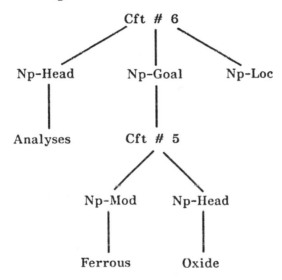

Fig 7. Semantic representation in tree format.

Case frame tokens are used by Syntax to expedite the building of syntactic structures consistent with Semantic hypotheses and to evaluate the ones it has built with respect to fulfilling or violating those hypotheses. Syntactically, there are only a few ways of structuring the set of cases shown in Fig. 5a. The head case must appear

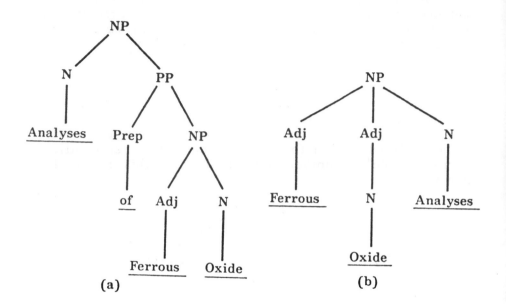

(a)

(b)

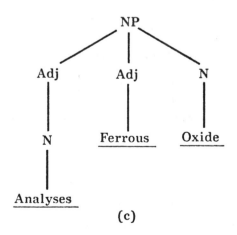

(c)

Fig 8. Syntactic structures.

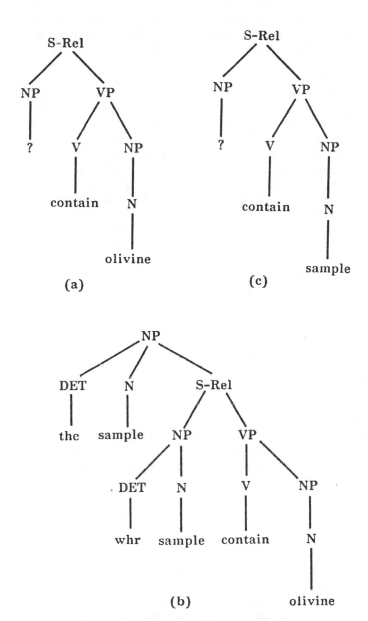

Fig 9. Structure for *the olivine containing sample.*

as the syntactic head and the object case must be realized
either in a prepositional phrase or relative clause or as an
adjectival modifier on the head. Thus, in Fig. 8, syntactic
structures (a) and (b) would confirm the semantic
hypotheses in Fig. 6, while (c), where "analyses" modifies
"oxide", would not and would therefore receive a lower
evaluation. Notice that the only difference between the
terminal strings of (a) and (c) is the presence of the
preposition "of". It takes only the presence of that small,
acoustically ambiguous word to allow Syntax to build a
structure consistent with Semantics' hypotheses. Knowing
this, Syntax and Semantics should be able to work together
to reconstruct and suggest to the word matcher these small
function words which make all the difference for correct
understanding.

The point is that Syntax should not make choices
randomly in places where Semantics has information that
can be used to order them. This is implemented via
Syntax's ability to ask questions of Semantics on the arcs
of the transition network grammar (Bates, 1974; Woods,
1970). For example, noun/present-participle/noun strings
may have the structure of a preposed relative clause like
"the olivine containing sample" (i.e., "the sample which
contains olivine") or a reduced relative clause like "the
sample containing olivine". (It may be that prosodies help
distinguish these two types of relative clauses in spoken
utterances, but it may also be the case that this additional
cue is not used if the phrase is already disambiguated by
semantics or context.)

In parsing the string "the olivine containing sample",
Syntax must chose whether the participle indicates a
preposed relative clause or a reduced one. If preposed,
"olivine containing" would have the structure shown in Fig.
9a, with "olivine" as object and subject unknown. This is
acceptable to Semantics, since olivine, a mineral, is a
possible rock constituent and hence containable. "Sample"
then becomes the head of the noun phrase and
simultaneously the subject of the preposed relative clause,
as shown in Fig. 9b. This Semantics accepts. Were the
word match one for "sulfur" instead of "sample", the final
structure -- "the sulfur which contains olivine" -- would be
semantically anomalous, and Semantics would advise Syntax

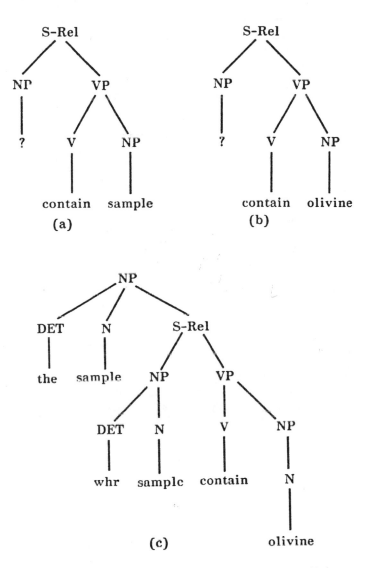

Fig 10. Structure for *the sample containing olivine.*

to look for another possible parsing. On the other hand, "sample containing", with "sample" as object (Fig. 9c), is semantically anomalous in the lunar rocks domain, so again Syntax would be advised to try again.

As a normal relative clause, "the olivine containing sample" has the intermediate structure shown in Fig. 10a, which is as bad as in 9c. Only "The sample containing olivine" is reasonable as a normal reduced relative clause (Figs. 10b and 10c). So Syntax's choice of parsing each string as a preposed or normal reduced relative clause will depend on its acceptability to Semantics.

VI. SUMMARY

Semantic knowledge is used in several ways to aid the general speech understanding task, as illustrated in SPEECHLIS: (1) It makes predictions local to a single utterance. (2) It collects sets of word matches which substantiate its hypotheses about the meaning of the utterance. (3) It checks the possible syntactic organizations of the word matches in order to confirm or discredit those hypotheses. These are done currently in SPEECHLIS using both a *semantic network* representing the concepts known in the domain and the words and multiword names available for expressing them, and also *case frames* which give further information about their surface and syntactic realization. Other mechanisms and procedures are being developed to extend the range and power of SPEECHLIS Semantics, to cover such things as cross-utterance predictions and the use of factual information and advice from Pragmatics in verifying its hypotheses.

Rollin Weeks, of the SDC Speech Understanding Project, has called automatic speech understanding the coming together of two unsolved problems -- speech and understanding. As such, it is an exciting and challenging field of investigation.

REFERENCES

Barnett, J. A vocal data management system. *IEEE Transactions on Audio and Electroacoustics*, 1973, *Au-21*(3), 185-188.

Bates, M. The use of syntax in a speech understanding system. *Proceedings of the IEEE Symposium on Speech Recognition*, Carnegie-Mellon University, April 1974.

Fillmore, C. The case for case. In Bach and Harms (Eds.), *Universals in linguistic theory*, Chicago, Ill.: Holt, 1968 1-90.

Forgie, J. W. Speech understanding systems, semiannual technical summary, 1 December 1973 - 31 May 1974. Lincoln Laboratory, Massachusetts Institute of Technology.

Fromkin, V. The non-anomalous nature of anomalous utterances. *Language*, 1971, *47*(1), 27-53.

Marcus, M. Wait-and-see strategies for parsing natural language. (MIT AI Working Paper 75), August 1974.

Neely, R. B. On the use of syntax and semantics in a speech understanding system, Department of Computer Science, Carnegie-Mellon University, May 1973.

Pollack, I., & Pickett, J. The intelligibility of excerpts from conversation, *Language and Speech*, 1964, *6*, 165-171.

Reddy, D. R., Erman, L. D., Fennell, R. D., & Neely, R. B. HEARSAY speech understanding system: An example of the recognition process. *Proceedings of the Third International Joint Conference on Artificial Intelligence*, Stanford University, August 1973, 185-193.

Riesbeck, C. K. Computational understanding: Analysis of sentences and context, unpublished doctoral dissertation, Stanford University, 1974. Reprinted in R. Schank (Ed.), *Conceptual Information Processing*, Amsterdam: North Holland, 1975.

Ritea, H. B. A voice-controlled data management system. *Proceedings of the IEEE Symposium on Speech Recognition*, Carnegie-Mellon University, April 1973.

Rovner, P., Nash-Webber, B., & Woods, W. A. Control concepts in a speech understanding system (BBN Report No. 2703). Cambridge, Mass.: Bolt Beranek and Newman (1973). (Also in *Proceedings of the IEEE Symposium on Speech Recognition*, Carnegie-Mellon University, April 1973.)

Sussman, G. Some aspects of medical diagnosis (MIT AI Working Paper 56), December 1973.

Walker, D. E. Speech understanding, computational linguistics and artificial intelligence (Technical Note 85). Menlo Park, California: Stanford Research Institute, Artificial Intelligence Laboratory, August 1973.

Walker, D. E. Speech understanding through syntactic and semantic Analysis. *Proceedings of the Third International Joint Conference on Artificial Intelligence*, Stanford University, August 1973, 208-215.

Wanner, E. Do we understand sentences from the outside-in or from the inside-out, *Daedelus*, Summer 1973, 163-183.

Winograd, T. PROGRAMMAR: A language for writing grammars (MIT AI Memo No. 181), November 1969.

Woods, W. A. Transition network grammars for natural language analysis. *Communications of the Association for Computing Machinery*, 1970, *13*(10), 591-606.

Woods, W. A., Motivation and overview of BBN SPEECHLIS: An experimental prototype for speech understanding research. *Proceeding of the IEEE Symposium on Speech Recognition*, Carnegie-Mellon University, April 1974.

Woods, W. A., Kaplan, R. M., & Nash-Webber, B. The lunar sciences natural language information system: Final report. (BBN Report No. 2388). Cambridge: Bolt Beranek and Newman, June 1972.

REASONING FROM INCOMPLETE KNOWLEDGE

Allan Collins
Eleanor H. Warnock
Nelleke Aiello
Mark L. Miller
Bolt Beranek and Newman
Cambridge, Massachusetts

I. INTRODUCTION

It does not trouble people much that their heads are full of incomplete, inconsistent, and uncertain information. With little trepidation they go about drawing rather doubtful conclusions from their tangled mass of knowledge, for the most part unaware of the tenuousness of their reasoning. The very tenuousness of the enterprise is bound up with the power it gives people to deal with a language and a world full of ambiguity and uncertainty.

We will describe this kind of human reasoning in terms of how a computer can be made to reason in the same illogical way. For this purpose we will use SCHOLAR (Carbonell, 1970a,b), a computer program whose knowledge about the world is stored in a semantic network structured like human memory (Collins & Quillian, 1972). One of SCHOLARs data bases is about geography, and people's knowledge about geography has the nice property, for our purposes, of being incomplete, inconsistent, and uncertain, so the examples and analysis will concern geography, but geography is only meant as a stand-in for everyman's knowledge about the world.

SCHOLARs aim in life is to teach people by carrying on a tutorial dialog with them (see Collins, Warnock, & Passafiume, 1975). Once upon a time, Socrates thought he could teach people to reason by such a tutorial method. We will attempt to show that a person can learn to infer at least some of what he does not know about geography by the Socratic method, and to show how a program like SCHOLAR might even play the role of Socrates with some finesse.

II. OPEN VERSUS CLOSED WORLDS

Recently Carbonell & Collins (1973) have stressed the distinction between open worlds, such as geography, where knowledge is incomplete, and closed worlds, such as the blocks world of Winograd (1972) or the lunar rocks catalog of Woods, Kaplan, & Nash-Webber (1972), where the complete set of objects and relations is known. The distinction is important, because many of the procedures and rules of inference which have been developed for dealing with closed worlds do not apply to open worlds.

The distinction between open and closed worlds arises in a variety of ways. For example, if there are no basaltic rocks stored in a closed data base, then it makes sense to say "no" to the question "Were any basaltic rocks brought back?"; but, if no volcanoes are stored in a data base for the U.S., it does not follow that the question "Are there any volcanoes in the U.S.?" should necessarily be answered "no". A more appropriate answer might be "I don't know." Furthermore, it makes sense to ask what the smallest block in a scene is, but it makes little sense to ask what is the shortest river or the least famous lawyer in the U.S. It would be an appropriate strategy for deciding how many blocks in a scene are red to consider each block and count how many are red; but it would not be an appropriate strategy to consider each person stored in a limited data base (such as humans have) in order to answer the question "How many people in the U.S. are over 30 years old?".

Within open worlds there are closed sets, however. For example, it may be possible to say how many states are on the Pacific, if they are all stored. Since closed sets are rare, it makes sense to mark the closed sets in memory rather than the open-ended sets. Then it is possible to apply closed-set strategies where the entire set is known.

The reason most sets are open is that most concepts are ill defined. One could plausibly argue that there is a smallest city in the U.S., if we agree on some arbitrary definition of a city (e.g., incorporation by a state), but to use Wittgenstein's (1953) example, there is no way to specify precisely what is and is not a game. Even if we were to agree on some definition, we would get into difficulty when we try to apply it to cases. Outside of mathematics and logic, most concepts are simply not susceptible to precise definition.

Where a concept is relatively well defined, like states in the U.S., we still may not know all the examples, and so we have to treat it as an open set. This means that the distinction between open and closed sets is not in the outside world, but rather in each person's head. Your closed set may be my open one.

We can illustrate some of the issues by considering Moldavia, since hardly anybody ever considers it, except perhaps Moldavians. Most adult Americans know all the states in the U.S., so they know that Moldavia is not a state. They may not be able to name all the states, but they have heard the states enough times that they have stored each of them as recognizably a state. They may even know either explicitly (to name) or implicitly (to recognize) all the countries in South America well enough to say Moldavia is not one of those.

The same distinction between explicit and implicit knowledge exists in SCHOLAR. The states would be stored implicitly if each appeared as an entry in the data base with an instance-of (superordinate) link to state. They would be stored explicitly if they were all stored as instances under U.S. states.

The same objects can be part of a closed set on some occasions and an open set on others. Even though a person (or SCHOLAR) may know all the countries in South America, he may not know all the countries in the world, so he may not be able to say whether Moldavia is a country or not, even though he can say it is not a country in South America. Similarly none of us really knows whether Moldavia is a city or town in the U.S., unless of course it is one. By restricting the set to, say, the major cities in the U.S., we can exclude Moldavia. Whether Des Moines is a major city in the U.S. may be debatable, but there is no way Moldavia can be. Words like "major" or "typical" (Lakoff, 1972) make it possible to restrict a set to exclude borderline cases, such as the likes of Moldavia.

What it takes for a computer system like SCHOLAR to discriminate between Moldavia and Des Moines are tags which indicate the relative importance of different cities (Carbonell & Collins, 1973; Collins, Warnock, & Passafiume, 1975). Suppose there is a particular data base configuration where a number of U.S. cities are stored, with Moldavia not one of them and Des Moines tagged to be of minor importance. The decision rule as to what are the major cities would be something like this: include those that are tagged as important, exclude any not stored, and any objects stored, but not clearly important, are excluded or hedged about, depending on their relative importance and the size of the set stored. In this way people can apply a modified closed-set strategy to deal with open sets.

This strategy is just one rabbit from a seemingly open-ended hat. People have more such tricks than we can see, much less understand. There are negative tricks, functional tricks, visual imagery tricks, inductive tricks, and undoubtedly many more that people use to circumvent the holes and uncertainties in their knowledge. These all lie outside the deductive logic of which the advocates of theorem-proving and the predicate calculus are so fond.

III. NEGATIVE INFERENCES

People do not store most things that are not true, for example, that Mexico has no king. Therefore, deciding that something is not true normally requires an inference. In a closed world, one can relegate whatever is not stored or not deducible from what is stored, to the dustbin of untruths, but in an open world, if one says "no" on that basis, then one will simply often be wrong. Therefore people use a variety of strategies to decide when to say "no", "probably not", "not really", or "I don't know".

Many of the strategies that people use to reach negative conclusions involve their functional knowledge, which we will discuss in the next section, or their visual knowledge, which we will not discuss in this chapter, but there are several strategies we have been modeling in SCHOLAR which are almost purely negative inferences.

A. Match Processes

Underlying the strategies people use to deal with their incomplete knowledge is a basic matching process for deciding whether any two concepts are the same or different. As we have argued elsewhere (Collins & Quillian, 1972), whether two concepts are treated as the same or different depends on the outcome of the matching process and the context in which it is called. The matching process pervades all language processing (Collins & Quillian, 1972) and we will refer to it repeatedly in the various inferential strategies described.

SCHOLAR does not mimic the human matching process perfectly (as described by Collins & Loftus, 1975) for reasons of computational efficiency, but it does capture the major aspects of that process. In SCHOLAR basic comparisons are handled by two processes called match-on-superordinate and match-on-superpart. Match-on-superordinate decides "Can X be a Y?" and match-on-superpart decides "Can X be part of Y", for any two elements X and Y in the data base.

Both processes are basically intersection searches (Quillian, 1968, Collins & Quillian, 1972). They trace along the superordinate or the superpart links from both X and Y until an intersection is found. When an intersection is found, they try to find a basis for saying "yes" or "no", and only if they fail do they conclude "I don't know". The

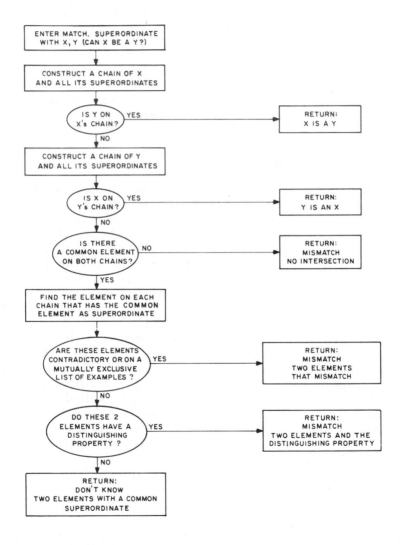

Fig. 1. Flowchart for match-on-superordinate.

flow chart in Fig. 1 shows the basic strategy of match-on-superordinate which is the simpler of the two processes.

We will explain how match-on-superordinate works in terms of its various outcomes in answering true-false questions. Examples (1) and (2) show different kinds of "yes" outcomes. (Expressions in lower case are inputs to SCHOLAR.) The first example involves a simple superordinate inference, and the second shows the outcome when the superordinate link is in the opposite direction from that which the question presupposes.

(1) Is Santiago a place?
     YES. MORE SPECIFICALLY, SANTIAGO IS A CITY.
(2) Is a city a capital?
     SOMETIMES. MORE GENERALLY, A CITY IS A PLACE.

If there is no common superordinate of X and Y, a reasonable response is "no". This may be the strategy people use to say "no" when two things are semantically unrelated. For example, when asked if a cafeteria is a dog, the reason they give for saying "no" is something like "A cafeteria is one sort of thing and a dog is another sort of thing". In example (3) the top-level superordinate for Surinam is place, which does not intersect with the chain for great man, and so the answer given is "no".

(3) Is Surinam a great man?
     NO. SURINAM IS A COUNTRY, NOT A GREAT MAN.

If there is an intersection found, and the two things that led to the intersection are marked as mutually exclusive, then match-on-superordinate concludes "no". People may use just such a strategy to say "no" when asked if a sparrow is a wren. They probably do not know any properties that are different between the two; they are just different kinds of birds. In example (4) Bolivia is a country and a capital is a city. City and country are mutually exclusive kinds of places, so match-on-superordinate concludes "no".

(4) Is Bolivia a capital?
    NO.   BOLIVIA   IS   A   COUNTRY,   NOT   A
    CAPITAL.

The next example illustrates the case where the two objects, in this case Recife and Montevideo, have a common superordinate, but are not mutually exclusive. People frequently give a distinguishing property, such as a difference in location, as a reason for saying that two things are not the same. This observation led to the test for a distinguishing property in match-on-superordinate. In example (5) Recife and Montevideo have the distinguishing property that they are located in different places, so the answer is "no". This difference is found by match-on-superpart.

(5) Is it true that Recife is the capital of Uruguay?
    NO.   RECIFE   IS   IN   BRAZIL,   NOT   IN
    URUGUAY.   THE   CAPITAL   OF   URUGUAY   IS
    MONTEVIDEO.

Example (6) shows the failure to find any basis for a distinction. A port can be a city and Sao Paulo is a city, and port and Sao Paulo are not stored as mutually exclusive. Nor are there any distinguishing properties between them. So there is no contradiction. This leads to a "don't know" outcome.

(6) Is Sao Paulo a port?
    I DON'T KNOW IF SAO PAULO IS A PORT,
    BUT IT IS A CITY.

When a "don't know" outcome occurs, it is appropriate to try a probabilistic inference, such as a lack-of-knowledge inference (Carbonell & Collins, 1973) or a functional inference. For example, it is possible to infer that Sao Paulo is probably not a port on the "lack-of-knowledge" basis that "I know other less important cities that are ports and so I would know about it, if it were true". Alternatively, the functional basis might be used that since it is not known to be on any major rivers or bodies of water, it must not be a port. These kinds of inferences will be discussed in more detail later.

B. Contradictions and the Uniqueness Assumption

Contradictions appear to be logically certain inferences, but people's contradictions turn out to be uncertain inferences, based on incomplete knowledge. We can illustrate the uncertainty of contradictions with examples from actual human dialogs. The following examples show the basic contradiction strategy people use.

(Q) Is Philadelphia in New Jersey?
(R) No. It's in Pennsylvania, but it's across the river from New Jersey.

(Q) Is Portuguese the language of Mexico?
(R) No. Spanish is the language of Mexico.

The contradiction strategy that emerges from these two examples (as well as others) depends on meeting four conditions. The conditions are specified in terms of what is found or not found in memory. In order to reach a contradiction to a query of the form "Is X in relation R to Y?" the memory search must meet the following conditions: (1) for all U that are found such that U R Y, U must be distinct from (i.e., mismatch) X, (2) for all V that are found such that X R V, V must be distinct from Y, (3) for all S that are found such that X S Y, S must be distinct from R, and (4) either the Us or Vs must be a complete set. The first three conditions can be satisfied by failure to find anything in memory (or by finding only some of the things there) but the completeness condition (4) cannot. These conditions are not at all obvious, and we will try to explain them in terms of one of the examples.

The way these conditions must have been satisfied in the Philadelphia example was as follows: (1) either he did not consider any places in New Jersey, or any he found (for example Newark or Camden) must have been distinct from Philadelphia, (2) the place he found Philadelphia to be in was Pennsylvania, and he must have found that to be distinct from New Jersey, (3) the relation he found between Philadelphia and New Jersey was "across the river from" and he must have found that to be distinct from "in", and (4) he assumed that Pennsylvania was the only place Philadelphia was in (i.e., that it was the complete set of locations for Philadelphia). Though the first of these conditions did not show up in the response in any form, it still must have been met. For example, if the respondent had known of a place called "East Philadelphia" in New Jersey, his memory search would probably have found it in this context (see Collins & Quillian, 1972) and he would have had to decide if it matched Philadelphia or not. In such a case he would probably have hedged his answer with "Well, there is an East Philadelphia in New Jersey."

The uncertainty in this kind of inference arises for two reasons. The most obvious reason is that the memory search hardly ever finds all the Us, Vs, and Ss which are relevant to the decision. We think that the search for Us, Vs, and Ss goes on in parallel, and the longer the question is considered, the more of them are found. What triggers a response is finding enough information to satisfy all the conditions. Sometimes, as we will see in a later example, a relevant piece of information is found after an initial answer has been given, and the person ends up changing his answer.

The other reason for the uncertainty lies in the default assumption of uniqueness that the respondents made in both examples, in order to meet the completeness condition (4). The uniqueness assumption has the form: If only one thing is found, then assume that it is the complete set. In the examples the respondents assumed that there is only one location for Philadelphia, and that Mexico has only one language, because these were the only values they found in memory. In fact, the assumption was wrong about Philadelphia; there is a Philadelphia, Mississippi, of civil rights fame, and there may well be a Philadelphia in New Jersey.

The completeness condition need not be satisfied for both Us and Vs. If one knows all the locations of Philadelphia, and they are all distinct from New Jersey, then it follows that Philadelphia cannot be in New Jersey. There is no indication that the respondent assumed, in the example, that he knew all the places in New Jersey, nor does that assumption ever appear to be the case.

In the examples below we will show how this basic strategy has been embedded in SCHOLAR. The first two are like the Philadelphia example in that the contradiction is based on information about the first object in the question. The mismatch occurs because Sao Paulo has only one location and there is a mismatch between it and the suggested location. In the first example the mismatch as to the location of Sao Paulo occurs on the basis of the country it is in, and in the second example on the basis of the region it is in. While the location is unique, it can be characterized in different ways, and a contradiction with any of those characterizations is grounds for saying "no".

(1) Is Sao Paulo in Chile?
     NO. SAO PAULO IS NOT IN CHILE, BUT IN BRAZIL.
(2) Is Sao Paulo in the Amazon jungle?
     NO. SAO PAULO IS NOT IN THE AMAZON JUNGLE, BUT IN THE BRAZILIAN HIGHLANDS.

In the third example Sao Paulo might have been rejected on the grounds that it is not among the cities in Chile, if that set were complete. Completeness can be satisfied either by making the uniqueness assumption or by marking the set as exhaustive. The uniqueness assumption does not apply, because there is more than one city stored. Furthermore, the set of cities is not marked as exhaustive. Therefore, the completeness condition is not satisfied for cities in Chile. But Sao Paulo is located in Brazil, not in Chile, and Brazil is distinct from Chile. That is to say the uniqueness assumption is applied to the location of Sao Paulo, just as in the first example. Example (4) is a variation on example (3). It illustrates the fact that the contradiction need not be based on a mismatch with respect to location. It can also be based on a mismatch with respect to the superordinate (i.e., cities are not mountains).

(3) WHAT ARE THE PRINCIPAL CITIES IN CHILE?
     Sao Paulo.
     NO. THE CORRECT ANSWER IS SANTIAGO, VALPARAISO, CONCEPCION, ANTOFAGASTA, ARICA, VINA DEL MAR, PUNTA ARENAS,

AND IQUIQUE. SAO PAULO IS NOT IN CHILE, BUT IN BRAZIL.
(4) Is Aconcagua a city in Peru?
NO. ACONCAGUA IS A MOUNTAIN, NOT A CITY. THE PRINCIPAL CITIES OF PERU ARE LIMA,   CALLAO,   AREQUIPA,   IQUITOS, TRUJILLO, CHICLAYO, AND CUZCO.

The next two examples show the two basic variants for finding a contradiction with respect to information about the second object in the question. In example (5) the strategy is like that of the person who said that Portuguese was not the language of Mexico. Because there is only one language stored with Uruguay, SCHOLAR assumes that Spanish is the only language. In example (6) there are two northern bordering countries stored for Argentina, but the set is marked as exhaustive, so the module can use this as a basis for saying "no".

(5) Is Portuguese the language of Uruguay?
NO. THE LANGUAGE OF URUGUAY IS SPANISH.
(6) Is Uruguay the northern bordering country of Argentina?"
NO.   THE   NORTHERN   BORDERING COUNTRIES OF ARGENTINA ARE PARAGUAY AND BOLIVIA. URUGUAY IS ONE OF THE NORTHEASTERN BORDERING COUNTRIES OF ARGENTINA.

In the first dialog excerpt we showed, the respondent pointed out that Philadelphia is across the river from New Jersey in order to explain the correct relation between the two objects in the question. Like the human, SCHOLAR will point out any other relation between the two objects it finds. In example (6) the search turned up the fact that Uruguay is the northeastern bordering country of Argentina, so like the human it added that information as clarification.

Example (7) shows the generality of the strategy for finding contradictions. Tropical refers to climate but nothing about climate is stored with Montevideo.

Montevideo is, however, part of Uruguay, and Uruguay's climate is temperate and maritime. There is a mismatch between tropical and temperate found by match-on-superordinate, and this is the basis for saying "no". This kind of nesting of inferences (here a part inference and a contradiction) is described by Carbonell & Collins (1973).

(7) Is Montevideo tropical?
    NO. MONTEVIDEO IS TEMPERATE.

We have argued above that people often use the uniqueness assumption as a default assumption to meet the completeness condition. This can be seen in the dialog following. The example is striking because it shows first how the uniqueness assumption operates to produce a contradiction, and then how it is overridden by finding more information.

(Q) Is Springfield in Kentucky?
(R) No.
(Q) Why do you say it's not in Kentucky?
(R) Because I know where Springfield is. It's in Massachusetts.
(Q) OK.
(R) There might be a Springfield in Kentucky. But I'm not really sure which one you're talking about.
(Q) Why didn't you bring that up when I asked you the question?
(R) Because I just assumed you were talking about Springfield, Massachusetts.

At the beginning of the dialog the respondent was first willing to say that Springfield was not in Kentucky, because it was in Massachusetts, but then she must have thought of another Springfield. (It is not uncommon to see people change their answers as they find more information in memory.) When she realized there was more than one Springfield and she did not know all of them, she gave a "don't know" kind of response. The reason she assumed the questioner was talking about Springfield Massachusetts, we would argue, is because that was the only Springfield she had thought of at first.

To see the extreme case of the uniqueness assumption, we recommend talking to a two-year-old. One two-year-old of our acquaintance, named Elizabeth, has been heard to respond to the accusation that she was a tease with the assertion "No, I'm a girl." This was striking because she did not know what a tease was. She knew she was a girl, and anything else had to be wrong. With age, people become less certain. It is hard to imagine that a man who was called a misogynist and who did not know what a misogynist was, would respond "No, I'm a man". It is absurd because adults have learned the multiplicity of things anyone can be. We suspect that people become less certain (and grow out of being "sophomoric") as they become more knowledgeable, because their greater knowledge leads to the storing of multiple values and prevents them from using the uniqueness assumption as a default assumption with the kind of abandon we see in our two-year-old.

The multiplicity of Elizabeth brings up the distinction between multiple values which are not equivalent, and sets (or lists) which are made up of equivalent elements. Instances such as Elizabeth or the Amazon only have one identity and one location at a time, in accord with the current physics of our world, but this identity or location can be described in a variety of ways. A person can be a two-year-old, a girl, and a tease; and the Amazon can be in South America, in Brazil and Peru, and even in the jungle. Though these multiple values look like sets, they behave differently from sets in some ways. One important distinction is that any one value will suffice in answer to a question or in making an inference. Thus for the location of the Amazon, it is appropriate to say simply that it is in South America or, alternatively, in Brazil and Peru. A set, on the other hand, is treated as a single element and should not be split into pieces. In the Amazon example, Brazil and Peru form a set, and so it is misleading to say simply that the location of the Amazon is Peru, just as it is misleading to say a zebra is black. It is not so bad, though, to say simply that the Amazon is in Brazil, because most of it is. When one or a few values within a set are predominant in importance, then they are often referred to as if they formed the complete set.

For the purpose of finding a contradiction, it is necessary to find a comparable element among the multiple values. Thus to decide if the Amazon is in the desert, it is appropriate to say "no" because it is in the jungle. On the other hand, the reason the Amazon is not in Argentina is because it is in Brazil and Peru. Failure to find a comparable element was the trap into which the uniqueness assumption led our two-year-old friend.

It would be possible to store explicitly the general knowledge which the uniqueness assumption makes implicit. For example, we might have stored as a fact about countries in general that they have only one capital and one language (unless otherwise indicated). The trouble with this approach is that, like knowledge of syntax, this kind of knowledge does not seem to be something that people usually know explicitly. For example, it comes as a surprise to discover that while countries have multiple products, mines usually have only one product. It is a generalization one has to make from all the mines one has encountered in the past. Thus such a scheme would lead to the storing of what appear to be little-known facts. While people may sometimes store such relationships explicitly, we would argue that in general this is implicit knowledge that is built into their inferential processes.

C. Lack-of-Knowledge Inferences

When they cannot find a contradiction, people sometimes fall back on what we call a lack-of-knowledge inference (Carbonell & Collins, 1973). This strategy can be seen in the following dialog excerpt:

(Q) Are there any other areas (the context is South America) where oil is found other than Venezuela?

(R) Not particularly. There is some oil offshore there, but in general oil comes from Venezuela. Venezuela is the only one that is making any money in oil.

In order to produce such a response we would argue the

respondent must have: (1) failed to find oil stored with any other South American country, (2) made the default assumption that most countries do not produce oil, (3) had oil stored as fairly important for Venezuela, and (4) had information stored about all the other countries that was less important than the importance of oil for Venezuela. On this basis the respondent concluded that if the other South American countries produced oil, he would know about it. He hedged his response in two ways: by saying "not particularly" rather than "no" and by saying Venezuela was the only one making money on oil. The hedges occur because at best this is an uncertain inference. In fact he was wrong; other South American countries were making money from oil, but not large amounts like Venezuela.

We can show how this strategy has been embedded in SCHOLAR with several examples. The lack-of-knowledge inference is called when there is a failure to find a contradiction. In such a case SCHOLAR compares the object in question to any similar objects that have the property in question. Suppose the question is about rubber in Guyana, as in the first example. SCHOLAR compares what it knows about Guyana and its agricultural products with what it knows about countries that produce rubber (i.e., Peru and Colombia). In particular, it compares the depth of its knowledge about agricultural products of Guyana to the depth at which rubber occurs for Peru and Colombia. The more SCHOLAR knows about Guyana, the more certain it is that it would know Guyana produced rubber if it were true. Similarly, the more important rubber is for those countries that produce rubber, the more certain it is that it would know if Guyana produced rubber. In this particular case it derived a moderate degree of certainty from the comparison.

(1) Is rubber an agricultural product of Guyana?
RUBBER IS AN AGRICULTURAL PRODUCT OF PERU AND COLOMBIA. I KNOW ENOUGH THAT I AM INCLINED TO BELIEVE THAT RUBBER IS NOT AN AGRICULTURAL PRODUCT OF GUYANA. THE AGRICULTURAL PRODUCTS OF GUYANA ARE SUGAR AND LUMBER.

Example (2) illustrates what happens when SCHOLAR cannot derive enough certainty from the comparison to say even a doubtful "no". In this case there were no minerals stored with the Chaco, so the comparison was based on the depth of its general knowledge about the Chaco. Because this comparison is less specific, less certainty can be derived from it. In this case, the result is a "don't know" response in which SCHOLAR points out the similar object it found where nitrates are a product.

(2) Are nitrates mineral products of the Chaco?
NITRATES ARE MINERALS OF THE ATACAMA DESERT. BUT I DON'T KNOW ENOUGH TO SAY THAT NITRATES ARE NOT MINERALS OF THE CHACO.

If there are no similar objects that have the property in question, then a "don't know" response is appropriate. In example (3) nothing about a humid climate is found for the other mountain ranges in South America, so SCHOLAR in its clumsy prose admits the extent of its ignorance.

(3) Is it true that the climate of the Andes is humid?"
I DON'T KNOW WHETHER HUMID IS A CLIMATE OF ANY MOUNTAIN RANGE IN SOUTH AMERICA.

The lack-of-knowledge inference involves reasoning from cases, like an induction. One kind of induction has the following form: if several objects in a class have a given property, then an object of the same class that one knows less about probably also has that property. In contrast, the lack-of-knowledge inference has the following form: if several objects in a class have a given property, then an object of the same class that one knows as much or more about probably does not have that property. Thus it turns out that the lack-of-knowledge inference is a kind of inverse of induction.

IV. FUNCTIONAL INFERENCES

People can often figure out what they do not know by reasoning from their knowledge about what it depends on. In geography people's knowledge about what depends on what is almost always incomplete. They often do not know all the functional determinants that influence a given variable such as climate, agricultural products, or population density. They are even less likely to know precisely how the different functional determinants (or factors) affect the variable, the values for all the functional determinants, or how the determinants interact. From the protocols we have looked at, however, people appear to reason like engineers. They make rough calculations using various default assumptions such as linearity, independence of factors, and normal values for unspecified factors. Then they make adjustments afterwards for perceived variations from these assumptions.

In this section, we will use excerpts from dialogs between a human teacher (T) and student (S) to illustrate these aspects of functional reasoning, as well as some of the different strategies people use in functional reasoning. All the excerpts are verbatim except for the last, which is reconstructed from notes.

The first example illustrates the form of people's functional knowledge; in particular, the temperature function and two of its functional determinants, latitude and ocean temperature. Here ocean temperature is treated as causing an adjustment of the temperature determined by latitude. What emerges from this and other examples is that temperature is regarded as a linear function of latitude, with adjustments for other factors like altitude, ocean temperature, and tree cover. These modifying factors are assumed not to affect the calculation unless they have unusual values. A person will never estimate the temperature of a place if he knows nothing about the latitude, but he he may make a rough calculation of the temperature if he knows the approximate latitude but not the other factors, by assuming normal values for the other factors (given no information to the contrary). This is true even though the variations in altitude (0 to 5 miles) affect temperature roughly as much as do variations in latitude. It is just that there is a clear default value near 0 in the distribution for altitude and none for latitude.

Example 1.
    (T) Is it very hot along the coast here (points to Peruvian Coast)?
    (S) I don't remember.
    (T) No. It turns out there's a very cold current coming up along the coast; and it bumps against Peru, and tends to make the coastal area cooler, although it's near the equator.

This example also illustrates another aspect of the storage of functional relationships: the distinction between general knowledge and specific knowledge. The general knowledge about temperature involves how it depends on various factors like latitude, altitude, and ocean temperature. The specific knowledge is information the tutor has stored about the fact that coastal Peru is cooler than comparable regions and about the cooling influence of the particular ocean current. The general knowledge is about "temperature" and the specific knowledge is about "the temperature of coastal Peru". A data base must, therefore, be able to have functional knowledge stored in both places, with pointers between the two indicating that the specific knowledge is a known instantiation of the general rule.

The second example shows a student answering both a "why" question (Why do they grow rice in Louisiana?) and a "why not" question (Why not in Oregon and Washington?). In answer to the first, the student mentioned only one functional determinant, the need for water. In the dialogs, people typically give only one or two reasons in answer to a "why" question, except when they have thought about the functional determinants previously. The reasons given are the matches found between the values stored for the particular place (in this case Louisiana) and the values required for the particular variable (in this case rice).

Example 2.
    (T) Where in North America do you think rice might be grown?

> (S) Louisiana.
> (T) Why there?
> (S) Places where there is a lot of water. I think rice requires the ability to selectively flood fields...
> (T) O.K. Do you think there's a lot of rice in, say, Washington and Oregon?
> (S) Aha, I don't think so.
> (T) Why?
> (S) There's a lot of water up there too, but there's two reasons. First the climate isn't conducive, and second I don't think the land is flat enough. You've got to have flat land so you can flood a lot of it, unless you terrace it.

In answering the "why not" question in Example (2), he mentioned three of the four determinants of rice growing. (He omitted soil fertility here, though it came up later.) A "why not" question in effect asks for any mismatches between the values required by rice and the values stored for the place in question. It is very unusual in a "why not" question to mention a functional determinant, such as rainfall, where the value stored for the place matches the value stored for rice. In this case it happened because water supply was primed (Collins & Loftus, 1975) by the previous discussion. That is in fact why the tutor picked Oregon, as we will discuss in the next section.

A mismatch on one factor is reason enough for not growing a given product, like rice. On the other hand, it is necessary to have matches on all the relevant determinants for a yes answer. Thus a correct answer to the first question about Louisiana would have mentioned all four factors.

Example 3.
> (T) They grow some wheat out in the plains. Do you have any idea why?
> (S) Boy, these are questions for a city boy, you know. For wheat, what do you need? You need fertile soil, and you need adequate rains, but not as much as you need for rice. You don't need a tropical climate for wheat. They grow

wheat way up in Canada with a shorter growing season. So you need fertile soil and some rain, and at least some section of time where the temperature doesn't go too far below freezing.

In this third example the same student named three of the four functional determinants to answer why they grow wheat in the Plains. (The fourth, terrain, is not so critical with wheat, so it is not surprising it was omitted.) Both wheat and rice growing occur over a range of temperature, so they are both threshold functions of temperature. For places on earth, rice growing has only one bound. There are places that are too cold, but none that are too hot. On the other hand, wheat growing has two bounds, though the student was only concerned with one in his response. There are places where it is too warm for wheat, as well as too cold for wheat. Agricultural products, and as we shall see, population density, are typically treated as threshold functions on the various functional determinants.

In his response he mentioned that wheat needed fertile soil and adequate rains, but not as much as you need for rice. In people's talk about such threshold functions as soil and rainfall, they only use fuzzy values such as fertile and adequate. We think it is important to be able to represent varying degrees of precision from the kind of values that appear in conversation to precise numbers, and to process either type as points on a continuum with a range of tolerance against which all matches or mismatches are evaluated.

Example 4.
(T) Do you think they might grow rice in Florida?
(S) Yeah, I guess they could, if there were an adequate fresh water supply. Certainly a nice, big, flat area.

Example 5.
(T) What kind of grains do you think they grow in Africa, and where, then? (Pause) Well, where would they grow rice if they grew it anywhere?
(S) If they grew it anywhere, I suppose they'd grow it it in the Nile region, and they'd grow it in

the tropics where there was an adequate terrain for it.

The fourth and fifth examples show how people can make calculations about a variable, if they know the functional determinants. In Example (4) the strategy for deciding whether rice is grown in Florida is to match Florida against all four functional determinants. He mentioned that it matched terrain, and he may have figured out that Florida would match on temperature. He voiced reservations about the match on water supply, so it was a doubtful match to his mind. If he had considered the requirement for fertile soil, he might have rejected Florida for this reason. It turns out that rice is in fact not grown in Florida. The fifth example shows a variation on the same strategy, where the student made a successful prediction. The procedure is to pick those places with the best overall match on all the functional determinants. In this case he was quite right about the Nile delta, and though he was more vague about the tropics, he was right as far as he went. These two examples show that functional knowledge gives people real predictive power, even though it is fallible.

Example 6.
    (S) Is the Chaco the cattle country? I know the cattle country is down there.
    (T) I think it's more sheep country. It's like western Texas, so in some sense I guess it's cattle country.

The sixth example shows a tutor making a functional analogy with respect to cattle raising. He thought of a region, western Texas, that matched the region in Argentina called the Chaco in terms of temperature, rainfall, and vegetation, the functional determinants of cattle raising. Since he knew that western Texas was cattle country he inferred that the Chaco might be as well. A negative functional analogy might have occurred if the student had asked whether the Chaco produced rubber. Since the Amazon jungle and Indonesia produce rubber, the tutor could have said "no" on the basis of the mismatch between the Chaco and those regions, with respect to temperature, rainfall, and vegetation.

Example 7.

    (T) How many piano tuners do you think there are in New York City?

    (S) Well there are 3 or 4 in New Haven, which has about 300,000 people. That's about one per 100,000. New York has about 7 million people, so that would make 70. I'll say 50 or 60.

Example 7 shows another variation on a functional analogy. The analogy is between New Haven, for which the requested value was known, and New York City. The functional dependence used is that the number of piano tuners depends on population size. Probably the respondent did not have this particular functional dependence stored, but generated it, because he knew that it is people who use pianos and because he could figure out the ratio of population sizes for the two cities. This is a particularly good example of the assumption of linearity (that the number of piano tuners increases linearly with population size) and a correction afterward of 15 to 30% downward for some deviation from the assumptions made. The adjustment might either have corrected for a perceived nonlinearity (that the number of piano tuners, like members of Congress, does not quite increase linearly with population size), or for a perceived difference between New Haven and New York on another functional determinant (e.g., New Haven may be more cultural on the average than New York). What should be emphasized is that either kind of correction is applied afterwards, and that the second kind entails an assumption of independence of the two factors, population size and culture.

These examples illustrate some of the various ways that people gain real inferential power from their imprecise knowledge about what depends on what. The next section shows how this kind of knowledge can be acquired.

V. LEARNING TO REASON

In Table I we show segments from a dialog on population density. The tutor was the first author, and this is one of several dialogs discussing functional interrelationships in geography. These dialogs had the character of an inquisition, complete with mental torture.

What is most apparent from the dialogs is that the students were learning a great deal. The dialog in Table I shows the most sophisticated of the students, and the student's learning in this dialog is particularly obvious. The similarity to a Socratic dialog is striking. What the students were learning was not so much facts about geography, but rather how to induce what is relevant and predict what is likely. In other words, they were learning to think like geographers.

A. What the Student Learns

As the student progressed in the dialog of Table 1, he accumulated a whole set of factors that affect population density. He learned, from dealing with a range of instances, what were the important determinants of population density. It is a process of inducing general knowledge from specific instances.

His early difficulty in answering the question about why density is high on the East Coast and his complaints about the meaninglessness of such questions, indicate that initially he had no general knowledge stored about the reasons for population density per se. He did have specific knowledge about the density in different places, and about some of the reasons for that density. For example, he knew that New York had a good harbor and its port facilities made it a center of population. He also knew that immigrants had poured into the East Coast, and often settled there, and that people are attracted to where the government is. But these were facts about New York and Washington which happened to be relevant to population density. It was knowledge stored with specific instances, not information stored with population density explicitly.

TABLE I.
Segments from a Tutor–Student Dialog about Population Density

T First, I am going to talk about population density. Where are the large densities in North America?

S In North America I would suppose the Northeast Corridor, Washington to Boston, would be the most densely populated area overall.

T Now, why do you suppose that is?

S Well, most of the air traffic passes back and forth between those places I believe. That's where you hear most of the problems about transportation.

T No. That's a true statement, and what I want to know is why.

S You want to know the proximate causes of it?

T Yes. The causes of why it is a true statement.

S Well, there are all those cities there, right?

T OK, why are the cities there?

S H'm. Well, you get to the question of why are cities located in certain places. Well, I guess for geographical and strategic reasons. New York is there, because it has the greatest natural harbor in the world, I hear. Ah. It was the place where our country was settled first and a lot of the immigration came here and a lot of the people tended to gravitate to those places. And political reasons, I suppose. Washington, being the capital of the country, attracts a lot of bureaucrats and professional people.

T OK. Where else is the density high?

S Well, working up from Washington, there's Baltimore.

T No, I mean what other areas. You named the Northeast.

S Other places that are dense would be the Chicago area.

T Why do you suppose that's a dense area?

S That seems like almost a meaningless question. Because there's lots of people there.

(Section omitted)

T Now, do you have any feeling for why regions in China are densely populated?

S Well, the proximate cause I suppose is lack of adequate birth control, and the population explosion.

T Why didn't that happen in Siberia?

S Yeah, there's probably a pretty strong interaction between the birth control practices which have only now become even possible and the climate and food supplies of an area. Political factors are in there too. I suppose it's possible there could be a population explosion in Siberia, but it would just take a hell of a long time for

it to get there. You don't really start to get a
population explosion unless there's an already adequate
population that keeps on growing inexorably. Then it
starts to get...

(Section omitted)

T   Why do you suppose Java has high population and the
other Indonesian islands have low population density?

(Section omitted)

S   Well, I would doubt there would be large cultural
differences between the islands, although I think some
parts are predominantly Hindu but most I think is
Moslem. Neither of those sects are particularly strong
on birth control. Climate differences aren't so severe.
Political - I think the seat of government is on Java.
T   But why is the seat of government there? Because the
people are there, right?
S   Yeah, maybe so. It doesn't make much sense to talk
about the availability of ports in an archipelago. There
must be thousands of them for the taking. Let's see,
there's climate and ports and politics and food supply.
Maybe the soil is different on Java than it is on some
of the other islands.
T   Hm hm, that's possible.
S   Maybe there's a difference in the political history of
that island and the others. There might have been.
Other islands could have been part of different political
organizations. I think they used to belong to the Dutch
- most of them did.
T   So did Java.
S   Yeah, most of it did, but maybe there was a famine on
some island that wiped out a proportion of the
population a few generations back. That's pretty
hypothetical. I would just suppose it had something to
do with politics and food supply. Not too much
difference in climate.
T   Yeah, I don't think the politics matter really. Yeah,
well, I might mention that Sumatra has a very
mountainous terrain.
S   Oh, the terrain. Yeah, and the other place would be
much flatter and better for rice growing and stuff.
Yeah.
T   You mentioned soil and you were hitting at it then.

In the course of the dialog, he derived the following factors in addition to those mentioned for the East Coast: foreign trade from the West Coast, birth control from China, climate and food supplies from the difference between Siberia and China, soil and terrain (the latter was brought up by the tutor) from Java, industrialization from Europe, minerals from South Africa, and seafood from West Africa. As he accumulated these factors they became explicitly stored as functional determinants of population density in general.

When he was confronted with the problem of why Java has a high population density and the rest of the Indonesian islands generally do not, he started going through the reasons he had accumulated to see if he could find a potential difference between Java and the other islands on the functional determinants of population density. What the inductive process had achieved up to this point was not so much that the number of facts stored had increased, but that the information had become stored with the general concept of population density. It was now available for processing with respect to Java. Because of this, the student in fact gained real inferential power. The answer about Java and the prediction shown in an earlier section about rice growing in the Nile delta are only two examples of how the accumulation of functional knowledge enables the student to reason in a generative way from incomplete knowledge.

Not only did the student accumulate reasons for population density that were already stored with specific instances, but he also generated some new reasons. For example, he may have known that China has a large population because of its lack of birth control, but he probably did not have stored the fact that climate and food supply were also reasons why China has a large population density. He brought these up when forced to compare China with Siberia and to say why one has a large population density and the other does not. This is another, separate aspect of induction.

This induction process involves finding what mismatching properties of China and Siberia can produce a difference in population density. Obviously the fact that Siberia is a region and China is a country will not account for the difference in population density, but the differences in climate and food-growing capability both can, so these are what the student mentioned. A little later, in discussing India, the student revealed the connection he found between climate and population density. His idea was that people will die of exposure if the climate is too cold. Other possible connections between climate and population density are that people are attracted to warmer climates (which is why Florida has a large population) and that climate affects food-growing. He probably did not find the former connection, but the latter connection was probably the basis for his bringing up food-growing.

We would argue that the connection between climate and population density is the result of an intersection process, like the one hypothesized by Quillian (1968, 1969). In the student's memory there must have been several different pieces of information: the fact that Siberia has a very cold climate, the fact that China has a moderate climate, the fact that prolonged exposure to cold leads to death, and the fact that death lowers population. Starting at Siberia, China and population density, the search had to find these four facts, which when taken together lead to a difference in population density between Siberia and China. Tying these facts together creates a new piece of information.

There were also a number of other things the students learned during the dialogs which we might enumerate briefly.

(a) They learned about second-order effects. When the student in the dialog shown added food supply to his list of things which affect population density, this made it possible to see that soil and terrain for Java, and the proximity of the ocean in West Africa, might affect food supply and thereby affect population density. Thus the induction that food supply is a factor permitted the further induction of these second-order factors.

(b) They learned about the multiplicity of reasons for any given fact. As we said, this student accumulated many reasons for population density. If he had been asked later why there is a large population density on the East Coast, we think he would have included such variables as climate, food supply, and industrialization. This is shown most clearly in Examples (3) and (4) in Section 4, where initially a student gave only one reason for rice growing in Louisiana, but later gave three reasons for wheat growing on the plains.

(c) They learned about feedback effects and interactions between different factors. This student pointed out (though not in the excerpt shown) a feedback effect that occurs with respect to capitals. A capital usually is located where the people are, but the fact that the capital is there tends thereafter to attract people to the area. Interaction between factors showed up in many cases in the dialogs. One such case was that ports are only important for trade if there is something to ship, which ties this factor to food supply and industrialization.

In summary, the students were learning by induction, and the dialogs showed two different aspects to this induction process. (1) The students were deriving new functional determinants by comparison of contrasting instances, and (2) they were accumulating general knowledge about functional determinants from the specific instances. In both cases, the process involves gathering specific pieces of knowledge scattered about in memory and storing them together in a new configuration where they are more available. This pulling together of old knowledge into new structures requires new interrelationships to be specified. It is the fundamental way new knowledge is created.

B. The Socratic Method of Teaching Geography

In the dialogs the tutor was following a strategy to force the student to think like a geographer. The agenda for the discussions was simply to discuss the functional determinants of geographical variables such as population density and agricultural products, for different places on the five major continents. There was no fixed set of questions to be asked, but there was an a priori determination to ask "where" questions, "why" questions, and "why not" questions. The "where" questions elicit what is stored as specific knowledge about the variable in question, or force a predictive calculation where nothing is stored directly. The "why" and "why not" questions elicit whatever reasons are stored explicitly, or force inductions. When the student could not answer a "where" question, the tutor usually provided the answers himself, and then asked the corresponding "why" or "why not" question.

During the dialogs, the tutor often picked a new place to ask about, and there was one strategy that he used systematically for picking a new place. This strategy in its most general form showed up near the beginning of the dialog on population density (but not in the fragment shown). The student kept mentioning ports as a reason for population density. So the tutor asked about Mexico because the population density occurs mainly away from the ocean and the ports. Then he picked Alaska because there are a lot of potential ports and very little population density. The two places were chosen to force the student to see that ports were neither necessary nor sufficient for population density. This strategy is in essence the "near miss" strategy which Winston (1973) found was necessary for a computer program to induce concepts from instances of those concepts.

The "near miss" strategy occurred throughout the dialogs. Other examples were the selection of Siberia after China in the dialog in Table I, and of Oregon and Washington in Example (2) in the Section 4. In the latter case, the student said that they grow rice in Louisiana because there was a lot of water there. This was an incomplete answer in that it omitted the warm climate, flat terrain, and fertile soil which are required for rice growing. So the tutor picked as a "near miss" a place which had the factor mentioned (i.e., a lot of water), but which did not grow rice. This was to make the student see that a lot of water was not enough. The tutor was precluded from picking a place where rice was grown and there was little water, because water is necessary for rice growing.

There were two other aspects of the tutor's strategy for picking places which emerged in the dialogs, particularly with the less sophisticated students. These are basic aspects of the strategy to force the student to learn from cases:

(a) The tutor picked well-known places with extreme values on important functional determinants. For example, in one dialog on population density, he asked why they have a low density in places like the Sahara, Tibet, and Alaska. These places were brought up to draw out from the student lack of water, mountainous terrain, and cold climate as factors causing low density. This is an effective strategy because it allows the student to derive functional determinants himself by dealing with cases where the relevant determinant is the most obvious explanation.

(b) The tutor picked different places with the same value on the functional variable (e.g., different places with high population density), where the value occurred for some of the same reasons and some different reasons. (This strategy parallels Winston's generalization cases.) For example, with population density, cases like Tibet and Alaska both involve cold climate, in one case because of the mountains and in the other case because of the latitude. This strategy is effective for two reasons. First, by repeating factors the tutor can see if the student can apply what he has learned about one place to another place. Second, by illustrating the different combinations of factors that lead to the same conditions, the student is forced to derive the most general form of the functional dependencies involved.

The major difficulty for a computer program to tutor by this method is for it to understand the answers by the student, but this is not, however, an insoluble problem, because the program does not have to understand the student very well. The program only has to see if the student has included those factors that the program knows to be relevant for the place in question. Teachers can read answers to questions on tests written in handwriting that they could not read otherwise. This is because they have a strong expectation as to what the answer should say. Similarly, in analyzing answers, the program can use its knowledge about what are functional determinants and what are possible values for any particular place and for any variable like population density or agricultural products. In this way, the program can build at least a partial understanding of what the student is saying or not saying, even when his answers are ungrammatical or incoherent. The beauty of the Socratic dialog is that a partial understanding is all that is necessary to guide further questioning. It is not altogether inhuman to carry on a conversation when you do not completely understand what the other guy is saying.

VI. CONCLUSION

What emerges from this view of human inferential processing is that people can often extract what they do not know explicitly from some forms of implicit knowledge by plausible but uncertain inferences. Cutting across the variety of strategies we have described, there are common aspects, in particular match-processing and the various default assumptions people make. We would argue that these are basic elements common to all human reasoning, and that they are overlaid with a variety of heuristic strategies people have learned in order to give reasonable answers in the face of their incomplete knowledge.

ACKNOWLEDGMENTS

This research was sponsored by the Personnel and Training Research Programs, Psychological Sciences Division, Office of Naval Research, under Contract No. N00014-71-C-0228, Contract Authority Identification Number, NR 154-330. Partial support also came from a fellowship to the first author from the John Simon Guggenheim Foundation. We would like to thank Susan M. Graesser who helped program the contradictions and Daniel G. Bobrow for his comments on a draft of this chapter.

REFERENCES

Carbonell, J. R.   Mixed-initiative man-computer instructional dialogues. Ph.D. dissertation. M.I.T., 1970(a).

Carbonell, J. R.   AI in CAI: An artificial intelligence approach to computer-aided instruction. *IEEE Transactions on Man-Machine Systems*, 1970(b), *MMS-11*, 190-202.

Carbonell, J. R., & Collins, A. M.   Natural semantics in artificial intelligence. *Proceedings of Third International Joint Conference on Artificial Intelligence*, 1973, 344-351. (Reprinted in the *American Journal of Computational Linguistics*, 1974, *1*, Mfc. 3.)

Collins, A. M., & Quillian, M. R.   How to make a language user. In E. Tulving and W. Donaldson (Eds.), *Organization of Memory*. New York: Academic Press, 1972.

Collins, A. M., & Loftus, E. F.   A spreading activation theory of semantic processing. *Psychological Review*, 1975, *82*(5).

Collins, A. M., Warnock, E. H., & Passafiume, J. J.   Analysis and synthesis of tutorial dialogs. In G. H. Bower (Ed.), *The psychology of learning and motivation* (Vol. 9). New York: Academic Press, 1975.

Lakoff, G.   Hedges: A study in meaning criteria and the logic of fuzzy concepts. *Papers from the Eighth Regional Meeting, Chicago Linguistics Society*. Chicago, 1972.

Quillian, M. R.   Semantic memory. In M. Minsky (Ed.), *Semantic information processing*. Cambridge: M.I.T. Press, 1968.

Quillian, M. R.   The teachable language comprehender. *Communications of the Association for Computing Machinery*, 1969, *12*, 459-475.

Winograd, T.   *Understanding natural language*. New York: Academic Press, 1972.

Winston, P. H.   Learning to identify toy block structures. In R. L. Solso (Ed.), *Contemporary issues in cognitive psychology: The Loyola Symposium*. New York: Halsted, 1973.

Woods, W. A., Kaplan, R. M., & Nash-Webber, B.   The lunar sciences natural language information system: Final report. (BBN Report No. 2388). Cambridge: Bolt Beranek and Newman, June 1972.

AUTHOR INDEX

Author Index

Author Index

SUBJECT INDEX

A

Access, 4, 22-25, 132, *see also* Retrieval, Search, Description
Actions, 3, 10-13, 142, 222, 279, 365, *see also* Verbs
ACTORs, 126, 132, 175, 187, 193
ACTs, 9, 243, 238-242, 266, 268-271, *see also* Verbs, Primitives
Agency, 281-282, 285, 293-295, 301
ALLOW, 215, 220, 230, 278, *see also* Enabling conditions, Causality
Ambiguity, 40, 108
Analogical representations, 2, 31-32, 105, 113
Analogy, 20, 132, 198, 404-405
Analysis versus synthesis, 04
Analysts, 106-126
Anaphoric reference, 76, 312, 314, 328-330, *see also* Reference
Anomaly, 38, 145, 147, 154, 159, 161, 168, 172-174, 177, 180-181, 377
Artificial intelligence, 41, 43, 50, 83, 152, 306
Associations, 44, 135, 200, 250 263, 355, *see also* Links
Attempts, 218-219, 222, *see also* Plans, Goals
Attention, 138-139, 145
Attribution theory, 50-53, 302
Augmented transition networks, (ATNs), 114, 325, 377, *see also* Parsing, Syntax
Automatic programming, 208
Axioms, 112, 186-187, 192, 199, 205

B

Backup, 12, 96, 194, 302
Beer, 89, 96

Beliefs

Beliefs, 17, 28, 36, 274-277, 283, 307
Beta structures, 136, 199, 203, *see also* MERLIN
Blocks-world, 106-110, 117, 162-175, 181, 189-90, 276, 384
Bottom-up processes, 109, 125, 140, 146, 234, 330-332
Boundary conditions, 337-338, 340
Branch points, 85-88

C

Canonical representation, 45-48, 119
Cases, 8, 44, 56-58, 135, 369, 379, *see also* Verbs
Causal chains, 24, 241, 246-253, 261-263, 265-268, 282, 336, 347
Causality, 212, 215, 241-242, 336, 347, *see also* Links, causal, Knowledge, functional
CAUSE, 215, 220, 226-228, *see also* Causality
Central mechanism, 146-148
Certainty, 80, 153, 362, 367, 392, 398-399
Changes of state, 222, 280, *see also* States
Choice, 30-32, 86
Classification, 198-199
Combinatorial explosion, 203
Complaint department, 171, 177-181
Completeness, 17, 391-397, *see also* Exhaustiveness
Concept-driven processes, 140-141, *see also* Hypothesis-driven processes
Concepts, 8-10, 85, 196-201, 353-355, 365-368, *see also* Nodes
Conceptual dependency, 36, 238-242, 247, 268-271, 276, 278
Conceptual structure, 117-120, 257-258
CONNIVER, 25

421

Subject Index

Consciousness, 89-90, 146-148, *see also* Self-awareness

Consequences, 222, 231, 340-342, *see also* Intended effects, Side effects

Consistency, 10, 30

Constraint satisfaction, 109, 125, 127, 338

Constraints, 69, 112, 335-336, 348, 351, 357
  between variables, 23
  global versus local, 109, *see also* Knowledge, global versus local

Content-addressable memory, 25-26

Context, 131-149, 200, 243, 352, 355, *see also* Descriptions, context-dependent
  nonstatic, 259-263, 327-330

Context frame, 340-342

Contingent knowledge structure, (CKS), 105-128

Continuity, 13-14

Contradictions, 29, 313, 391-397

Control structure, 105, 112, 147, 177, 208, 362

Correction, 154, 158, 163

Counterexamples, 318, 341-342, 347

Criteriality, 28, 72

D

Data-driven processes, 109, 274, *see also* Event-driven processes, Demons

Data-limited processes, 141-143

Debugging, 192, 208

Decision processes, 86-88, 91, 97

Declarative knowledge, 15, 185-195, 201, 206, 208, *see also* Knowledge

Decomposability, 191-195

Deduction, 45, 112, 121, 186-193, 200, 205, 334

Deep structures, 44

Default assumptions, 198, 398, 400, 405, 414

Default values, 15, 155-157, 314, 400-401 *see also* Normality conditions

Defining versus asserted properties, 53, 58

Definite versus indefinite entities. 65-68

DELTACTS, 282-303, 306, *See also* Change of state

Demons, 93, 98, 124, *see also* Data-driven processes

Depth of processing, 145

Descriptions, 153, 156-157, 163-165, 198-202
  context-dependent, 25

Dialog, 406-413

Disabling conditions, 277, 279, 303, *see also* Enabling conditions

Discrepancies, 144

Domain and range, 4, *see also* Range of values

Dremes, 275

E

Education, *see* Teaching

Efficiency, 31, 83

Electronics, 176-177, 312-313, 334-335, 338

Ellipsis, 312, 314, 326-328, 355, *see also* Reference

Enabling conditions, 215, 241, 243-252, 281, 282-306

Episodes, 213-215, 222, 240-253, 264-267, *see also* Stories

Episodic memory, 256, 263, 306-307

Epistomology, 187, 192

Errors, 137

Established conditions, 244-246

Event-driven processes, 140-142, 145, *see also* Data-driven processes

Events, 144, 214-215, 355, 365

Executives, 87-103
  bookkeeping, 96-97

Exhaustiveness, 7, *see also* Completeness

Expectations, 108, 112, 144, 154, 171, 179, 351, 355

Experts, 104-106, 145, *see also* Procedural Specialists, Specialists

F

Feedback loops, 94, 111-113

Fillers, 69, *see also* Slots, Cases

Forgetting, 241, *see also* Memory

Frame problem, 11, 93, 96, 99, *see also* Temporal organization

Frames, 24, 132, 151-154, 185, 195-211, 255, 307
  frame axiom, 189
  transformation, 154, 172-174, 181

Fuzziness, 330, 403

Subject Index

G

GSP natural language system, 127
General Problem Solver, (GPS), 195
Generalization, 18, 413
  hierarchy, 196-201, 205
Generate and test paradigm, 333, 342
Global interpretations, 109
Global versus local control, 92, *see also*
    Knowledge, global versus local con-
    straints, global versus local
Goal synthesis, 114
Goal-oriented procedures, 121-123
Goals, 89-91, 96, 146, 301, 305-306, *see*
    also Plans
Graceful degradation, 137, 141, 159

H

HACKER, 19, 195, 208
Having, 280, 288-289, *see also* Possession
HEARSAY, 357
Hedges, 386, 398-399
Heuristic knowledge, 190-191, 205
Hierarchical organization
  of frames, 181-182, 196-214
  of processes, 84, 145-148
Hill-climbing techniques, 26, 344
Hypothesis, 97, 165-170, 174, 176-182,
    313, 318-323, 361, 371-376, *see also*
    Predictions
  evaluation, 313, 318-320, 333, 329-332,
    347-348
  generation, 313, 322-323, 333, 342-345, 348
Hypothesis-driven processes, 108, 274, *see*
    also Concept-driven processes
Hypothetical questions, 321, 323, 332-339,
    347, *see also* What-if questions

I

If-added/if-needed procedures, 123, *see*
    also Antecedent/consequent theorems
Implication, 47, *see also* Inference
IMPs (important elements), 156, 199-209
Indexing, 109, *see also* Access

Inducements, 294-296, *see also* Agency
Induction, 20, 85, 91, 159, 208, 399,
    406-408
Inference, 4, 16-20, 23, 29, 37, 41-42,
    45, 54, 105, 120, 261-262, 273,
    276, 302, 312-313, 332-348, *see*
    also Reasoning, Deduction, Induction,
    Contradiction, Analogy
  computational, 16, 19-21
  from examples, 22
  formal, 16-19
  functional, 20, 390, 400-405, *see also*
    Causality
  lack-of-knowledge, 390, 397-9
  meta, 16, 19-21, *see also* Analogy,
    Inference, Lack-of-knowledge
  negative, 386-398
  preferred, 17
  rules of, 112, 120, 192, 208, 301, *see*
    also Modus Ponens
Inheritance of properties, 197, *see also*
    Links, ISA
INITIATE, 214-215, 222, 229, *see also*
    Causality
Instantiation, 44, 69, 117, 125, 153,
    157, 160, 163-165, 175-176, 260,
    344, 365-367, 371
Instruction, *see* Teaching
Instrumentality, 227, 279-280, 284-297
Intended effects, 279-282, *see also* Side
    effects
Intension versus extension, 48
Intensional objects, 37, 50, 52, 59, 67-69,
    see also Nodes, intensional
Intentions, 278-280, 283, *see also* Plans,
    Goals
Interference, 147
Interrupts, 94
Intersection, 25, 387-389, 410

K

Knowing, 280, 290-291
Knowledge, *see* Representation, World
    Knowledge, Heuristic knowledge,
    Fuzziness, Model of the world
  contingent, 86
  domain-specific, 190-195

423

Subject Index

O

Observations, 177-180

Open versus closed worlds, 384

Operational correspondence, 4, *see also* Model of the world

Operations, 3, 119-121

Operators, 73-80, 195

Organization of memory, 239, 255-68, *see also* Hierarchical organization

P

Paraphrase, 9-10, 45-48, 136, 238-242, 249-254, *see also* Conceptual dependency, Primitives

Parsing, 26, 42, 109, 113, 125, 237, 240, 325-329, 362, 370, 377, *see also* Syntax, Semantic grammar, Augmented transition networks

Pattern-directed invocation, 194

Pattern-recognition, 163-170

Perception, 108, 125, 134-135, 145, 152, *see also* Recognition

Perturbations, 154, 159-161

Philosophy, 38-42, 48, *see also* Epistomology

PLANNER, 20, 123, 205, 338

Plans, 12, 29, 143, 190, 219, 222-223, 230, 278-292, *see also* Goals

Possession, 288-289

Power, 281, 296-298

Predicate calculus, 15, 17, 23, 44, 114, 121, 188, 386

Predicates, 51-56, 69-71, 77-79, 93, 199, 212, 239, *see also* Relations, Links

Predictions, 145, 153, 159-161, 163, 169-171, 326, 358, 363, 369, 371-376, *see also* Hypotheses

Prevention, 303-304

Primitive operations, 113, 116, 135-136, 278-279

Primitives, 9, 278, *see also* ACTs, Conceptual dependency

Problem solving, 117, 301

Procedural attachment, 203-206

Procedural knowledge, 15, 85, 185-195, 201, 206, 208

Procedural specialists, 110, 166-171, 175-177, 312, 322, 326-327, 329-330, 334-348, *see also* Specialists, Experts

Procedures, 106, 205-208

Production systems, 194, 342

Pronomial reference, 24, *see also* Reference, Anaphoric reference

Proof procedures, 186-187, 192, *see also* Predicate calculus

Propositional representation, 2, 31-32

Proximity, 282-288, *see also* Location

Psychology, 13, 41, 50, 84, 138-140, 144-145, *see also* Memory, Perception, Recognition

Purpose, 90, 369, *see also* Goals, Plans

Q

Qualities, 279, 281, 291-293

Quantification, 15-18, 36, 59, 68, 71-79, 331, *see also* Predicate calculus

Query evaluators, 111-113

Query language, 113

Question-answering, 2-3, 18-22, 30, 73, 110, 117, 120, 286, 311-312, 326, 333-339, 386, *see also* Inference, Reasoning, Knowledge

R

Range of values, 14-93, 154, 180-181, 345, 403, *see also* Variability, Domain, Range

Rationalizations, 347

Reasoning, 152, 198-199, 347, *see also* Inference

common sense, 12, 383-414

by examples, 338, 348

Recognition, 97, 142, 153-183, *see also* Perception, Visual memory, Memory

Redundancy, 48, 111, 206-207, 333, 345-346

Reference, 42, 132, 136-137, *see also* Anaphoric reference

Relations, 6, 41, 365, *see also* Predicates, Links

binary, 55

more than two arguments, 55-58

Relative clauses, 8, 36, 60-65, 331, 377

indefinite, 65-68

425

Subject Index

shared subpart fallacy, 60-62
   transient process account, 73
Relaxation techniques, 26, 125
Remembering, 144-145, *see also* Memory
Representation, 2-33, 250-253, *see also*
   Frames, Semantic networks, Knowledge, Procedural representation,
   Declarative representation, Visual
   representation, Multiple representations, Analogical representations,
   Propositional representations
   ad-hoc, 121
   of complex sentences, 65
   of episodes, 240-253, 258, 264-267
   general purpose, 121
   intensional, 67-69, 73
   levels of, 120
   logical adequacy, 45, 75, 79
   of plans, 278-311, *see also* Goals
   semantic, 42-46, 67, 237-242, *see also*
      Semantics
   of stories, 216-218, *see also* Stories
   syntactic, 42, *see also* Syntax
Resolution, 17, *see also* Theorem provers
Resource-limited processes, 141-143
Resources
   expenditures, 90
   conflicts, 91-92, 97
Retrieval, 22-26, 41, 47, 106, 114, 132-
   135, 138, 146-147, 359, *see also*
   Access, Search, Inference
Robots, 189-190
Roles, 265-267

S

SAD SAM, 47-48
SCA model, 103-128
SCHOLAR, 21-22, 28, 311, 383-398
Scenarios, 23
Scheduling, 29, 144
Schemata, 131-149
Scope, 71, 76-79, 203, *see also* Quantification
Scripts, 125, 264-267, 275-278, 284, 306
Search, 12, 18, 24, 46, 116, 198, *see also*
   Access, Retrieval
Selectional restriction, 42
Self-awareness, 4, 190, *see also* Consciousness
Semantic features, 42
Semantic grammar, 312, 324-332, 358

Semantic interpretation, 39-40, 62-65, 71-
   72, 114-116, *see also* Understanding
   language
Semantic memory, 36, 255, 263, 292
Semantic networks, 6, 8, 15-16, 22-25,
   31, 44-82, 106, 111, 136, 186, 199,
   312, 321, 325, 363-367, 379,
   383-384, *see also* Links, Nodes
   Concepts
Semantics, 351-380, *see also* Knowledge,
   Representation
   conceptual, 354-356, *see also* Concepts,
      Knowledge, World knowledge
   lexical, 256, 353-354, *see also* Words,
      Lexical memory
   procedural, 39-40, 42
   of programming languages, 43
Serial versus parallel processing, 94-95
Setting, 213-214, 222
Short-term memory, 147, 194
SHRDLU, 279, 325, *see also* Blocks world
Side effects, 192, 288, 335, *see also* Plans,
   Goals, Consequence
Similarity, 134, 238
Simulation, 105, 111-113, 312, 333-339,
   see also Procedural knowledge
SIR, 50
Skolem functions, 75-78
Slots, 67, 70, 125, 199, 238, *see also*
   Frames, Fillers
SMALLTALK, 175
Social actions, 280-285
Socratic method, 411-413, *see also*
   Teaching
Soul, 88, *see also* Demon
SOPHIE, 31, 110, 311-349
Specialists, 104, *see also* Procedural
   specialists, Experts
SPEECHLIS, 358
Spheres of influence, 88-90, 92
State representation generator, 111
State representation scheme, 110
States, 2, 90, 214, 222, 243-245, 268,
   271, 280, 284, 305-306, *see also*
   Change of state, World versus
   knowledge
Statistical information, 98-99
Storage, 22, 24, 31, 188, *see also* Knowledge,
   Learning, Economy
Stories, 211, 240-253, 267, 277-278, *see*

426

Subject Index